Predictive Models for Water Quality in Distribution Systems

Awwa
Research
Foundation

About the Awwa Research Foundation

The Awwa Research Foundation (AwwaRF) is a member-supported, international, nonprofit organization that sponsors research to enable water utilities, public health agencies, and other professionals to provide safe and affordable drinking water to consumers.

The Foundation's mission is to advance the science of water to improve the quality of life. To achieve this mission, the Foundation sponsors studies on all aspects of drinking water, including supply and resources, treatment, monitoring and analysis, distribution, management, and health effects. Funding for research is provided primarily by subscription payments from approximately 1,000 utilities, consulting firms, and manufacturers in North America and abroad. Additional funding comes from collaborative partnerships with other national and international organizations, allowing for resources to be leveraged, expertise to be shared, and broad-based knowledge to be developed and disseminated. Government funding serves as a third source of research dollars.

From its headquarters in Denver, Colorado, the Foundation's staff directs and supports the efforts of more than 800 volunteers who serve on the board of trustees and various committees. These volunteers represent many facets of the water industry, and contribute their expertise to select and monitor research studies that benefit the entire drinking water community.

The results of research are disseminated through a number of channels, including reports, the Web site, conferences, and periodicals.

For subscribers, the Foundation serves as a cooperative program in which water suppliers unite to pool their resources. By applying Foundation research findings, these water suppliers can save substantial costs and stay on the leading edge of drinking water science and technology. Since its inception, AwwaRF has supplied the water community with more than $300 million in applied research.

More information about the Foundation and how to become a subscriber is available on the Web at **www.awwarf.org**.

Predictive Models for Water Quality in Distribution Systems

Prepared by:
James Powell, **Jonathan Clement**, and **Malcolm Brandt**
Black & Veatch, 69 London Road, Redhill, Surrey RH1 1LQ, UK

Rob Casey and **David Holt**
Thames Water Plc, Spencer House, Manor Farm Road, Reading, Berkshire RG2 0JN, UK

Walter Grayman
W.M. Grayman Consulting Engineer, Cincinnati, Ohio 45215

and

Mark LeChevallier
American Water, 1025 Laurel Oak Road, Voorhees, NJ 08043

Sponsored by:
Awwa Research Foundation
6666 West Quincy Avenue, Denver, CO 80235-3098

Published by:

Awwa Research Foundation
Advancing the Science of Water®

IWA Publishing

DISCLAIMER

This study was funded by the Awwa Research Foundation (AwwaRF). AwwaRF assumes no responsibility for the content of the research study reported in this publication or for the opinions or statements of fact expressed in the report. The mention of trade names for commercial products does not represent or imply the approval or endorsement of AwwaRF. This report is presented solely for informational purposes.

Published by IWA Publishing, Alliance House, 12 Caxton Street, London SW1H 0QS, UK

Telephone: +44 (0) 20 7654 5500; Fax: +44 (0) 20 7654 5555; Email: publications@iwap.co.uk;
Web: **www.iwapublishing.com**

AwwaRF report number 91023F. Originally published by AwwaRF for its subscribers in 2004.
IWA Publishing version published 2005.

ISBN 978-1-84339-913-1

CONTENTS

TABLES

FIGURES

FOREWORD

The Awwa Research Foundation is a nonprofit corporation that is dedicated to the implementation of a research effort to help utilities respond to regulatory requirements and traditional high-priority concerns of the industry. The research agenda is developed through a process of consultation with subscribers and drinking water professionals. Under the umbrella of a Strategic Research Plan, the Research Advisory Council prioritizes the suggested projects based upon current and future needs, applicability, and past work; the recommendations are forwarded to the Board of Trustees for final selection. The foundation also sponsors research projects through an unsolicited proposal process; the Collaborative Research, Research Applications, and Tailored Collaboration programs; and various joint research efforts with organizations such as the U.S. Environmental Protection Agency, the U.S. Bureau of Reclamation, and the Association of California Water Agencies.

This publication is a result of one of these sponsored studies, and it is hoped that its findings will be applied in communities throughout the world. The following report serves not only as a means of communicating the results of the water industry's centralized research program but also as a tool to enlist the further support of the nonmember utilities and individuals.

Projects are managed closely from their inception to the final report by the foundation's staff and large cadre of volunteers who willingly contribute their time and expertise. The foundation serves a planning and management function and awards contracts to other institutions such as water utilities, universities, and engineering firms. The funding for this research effort comes primarily from the Subscription Program, through which water utilities subscribe to the research program and make an annual payment proportionate to the volume of water they deliver and consultants and manufacturers subscribe based on their annual billings. The program offers a cost-effective and fair method for funding research in the public interest.

A broad spectrum of water supply issues is addressed by the foundation's research agenda: resources, treatment and operations, distribution and storage, water quality and analysis, toxicology, economics, and management. The ultimate purpose of the coordinated effort is to assist water suppliers to provide the highest possible quality of water economically and reliably. The true benefits are realized when the results are implemented at the utility level. The foundation's trustees are pleased to offer this publication as a contribution toward that end.

Walter J. Bishop
Chair, Board of Trustees
Awwa Research Foundation

James F. Manwaring, P.E.
Executive Director
Awwa Research Foundation

FOREWORD

ACKNOWLEDGMENTS

The authors are grateful to Awwa Research Foundation (AwwaRF) and the participating utilities listed below. The help and support from Misha Hansan and Jian Zhang in their role as AwwaRF project managers was greatly appreciated. Particular thanks are given to the Project Advisory Committee (PAC) members for their time in reviewing the report: Polly Boissevain of Camp Dresser & McKee (CDM), Bill Kirkpatrick of East Bay Municipal Utility District (EBMUD), and Dan Pedersen of The City of Austin Water Utility.

Participating Utilities

- Greater Cincinnati Water Works
- City of Grand Rapids
- Tampa Bay Water

Workshop Attendees

The conclusions of this report and the research briefs proposed were largely identified at a workshop attended by a wide range of experts from academia, consultants, and water companies who are listed below.

Role	*Name*	*Organization*
AwwaRF project manager	Misha Hansan	AwwaRF
PAC	Polly Boissevain	CDM
	Bill Kirkpatrick	EBMUD
Project team	Malcolm Brandt	Black & Veatch
	Rob Casey	Thames Water Utilities
	Jonathan Clement	Black & Veatch
	David Holt	Thames Water Utilities
	Mark LeChevallier	American Water
	James Powell	Black & Veatch
Academic experts	Anne Camper	University of Montana
	Richard Valentine	University of Iowa
	Don Wood	Kentucky University
Industry experts	Robert Clark	Independent Consultant
	Walter Grayman	Independent Consultant
	Darren Lytle	USEPA
Participating utilities	David Hartman	Greater Cincinnati Water Works
	John Wierenga	City of Grand Rapids

EXECUTIVE SUMMARY

RESEARCH OBJECTIVES

The objective of this research project was to characterize the current state of predictive distribution system water quality models and to identify research needs for their improvement. The report is intended to steer future research and to act as a general reference on water quality modeling. The topic is addressed under three subject areas:

- Modeling chemical and aesthetic water quality
- Modeling microbial water quality
- Applying models to distribution networks

The report combines a literature review with the practical experience of the project team. The content of a draft report was discussed at an international workshop, attended by academics, engineers, scientists and hydraulic modelers with the objective of agreeing specific research needs necessary to improve predictive modeling for water quality in distribution systems. The conclusions of this report are derived from the workshop and form the basis of eleven specific research briefs that have been submitted to Awwa Research Foundation for consideration of funding.

WHAT ARE MODELS?

A model is a system that is used to understand the behavior of another system to which it is analogous in some way; generally it is simpler than the system being represented. Models come in a variety of forms including:

- *Conceptual models:* Diagrams, flow charts, or text which describes the system's behavior
- *Physical models:* Physical representations often at a reduced scale such as pipe rigs, scale models of tanks or jar tests
- *Numerical models:* Mathematical equations which define the behavior of a system

Models can be used to predict:

- System behavior at a location where it is difficult or costly to measure directly
- System behavior at another time under a changed environment
- The likely impact of altering the structure of the system

MODELING WATER QUALITY PROCESSES

Researchers often focus on modeling the individual processes which control water quality rather than fully modeling water quality throughout distribution networks. These 'process models' simulate the behavior of an isolated vessel of water and so ignore the complexities caused by flow patterns and the variations in the properties of pipes and other components which occur in real distribution networks. Brief details on the modeling of key water quality parameters follow

Corrosion and Metal Release

The metals with the greatest significance for water quality are iron, lead, and copper. The mechanisms and constraints governing the corrosion and solubility of lead and copper are reasonably well understood and can be modeled using numerical solubility models and physical pipe loops.

Iron corrosion can cause severe discoloration, production of sediment, and may promote microbial growth. Much of the recent research has focused on simple conceptual models that describe how iron is released from unlined cast-iron pipes. However, the release mechanism is complex and depends on a variety of factors including the precise mixture of compounds in the solid phase. Pipe loops can be used to investigate iron release, but they are considerably larger than those used to investigate copper or lead solubility because they need to simulate the behavior of distribution pipes rather than domestic plumbing.

The mechanisms of corrosion and metal release need to be researched to understand the compounds that reside on the internal surface of corroded iron pipes. With this information numerical models could be developed to help predict treatment changes for controlling discoloration.

Discoloration

Discoloration involves the production, accumulation, disturbance and transportation of fine particles within distribution. Several issues currently restrict the ability to build detailed models of discoloration:

- There is a lack of numerical models for the rate of particle production by iron release from corroding mains, precipitation reactions, or microbial growth
- Sediment transport has been widely researched over the last 50 years and a number of different models have been developed for canals, rivers, estuaries, slurry transportation, or sewage. In these applications, the sediment concentrations are higher and the particles often larger than is common in water distribution mains. Furthermore, these existing models do not fully account for the adhesion to and erosion from pipe surfaces
- Sediment transport is controlled by the properties of the sediment, in particular the size, density, and its concentration. There is a need for research on the characterization sediment properties in water distribution. This needs to consider the change in properties as sediment passes through the network due to chemical reactions or the attachment of microorganisms.

Disinfectant Decay

Free Chlorine

The rate of free chlorine decay in bulk water is controlled by both the concentration of chlorine and the concentration of the organic compounds with which it reacts. Recent research has shown that a second order model may be more robust than the frequently used first order model, because it can take account of the concentration of the organic reactant. In reality water comprises a complex matrix of chemicals with different reactivity. The impracticality of modeling all these individual reactions will always be a constraint when modeling decay.

The rate of wall decay is predominantly controlled by the pipe material and condition. Unlined iron pipe causes rapid decay and where present is normally the limiting factor in disinfectant loss in distribution systems. Although the relative magnitude of wall decay rate for different pipe materials is known, there can be considerable variability within and between materials.

Researchers have noted a tendency for an increased rate of decay with increased velocity. The reason for this is not understood, but it could be related to the mechanism by which chlorine transfers from the bulk flow to the pipe wall or the disturbance of sediment which subsequently reacts with the chlorine.

Chloramines

Chloramines are often modeled assuming first order decay but using lower decay coefficients than for free chlorine. However, the decay process is significantly different to that for free chlorine. Monochloramine is inherently unstable and will decompose producing free ammonia. This autodecomposition is strongly controlled by pH which is not accounted for by the first order decay model. Specific chloramine bulk water decay models have been developed which relate the model input variables to measurable parameters and known rate constants.

There is a need for increased understanding of monochloramine loss due to the pipe wall effects, with the aim of predicting how this will vary within and between distribution networks with apparently similar pipe work.

Disinfection By-Products

Trihalomethanes (THM) and haloacetic acids (HAA) are formed by the reaction of chlorine or chloramine with natural organic matter (NOM). The nature of the NOM and its reactivity with chlorine is highly diverse. As a result THM and HAAs are hard to model and tend to be based on empirical disinfection by-product (DBP) formation rate data for specific water conditions. These empirical models often use total organic carbon (TOC) as a model input to represent all NOM, but this does not account for the reactivity of the specific mixture of NOM present in a water. Mechanistic models would enhance the understanding of how different factors control DBP formation, but they require improved methods for NOM characterization.

Most modeling of DBPs has focused on total THM formation due to reactions with free chlorine. Although some of these models have also been applied to HAA formation, only one model has been identified in this review which was specifically developed for HAAs. This model simulates HAA formation within treatment plants.

In some cases organics bound to iron pipe deposits can cause an increase in DBPs, while adsorption onto iron pipe deposits can cause a decrease in concentration. The models do not account for these pipe wall effects which can be particularly important for systems with a significant amount of iron pipe. In such systems, predicting DBPs could be difficult.

The values of the coefficients used in the different models vary depending on the THM of concern and will, to a certain extent, be site specific. It is unlikely that they will account for interactions with the pipe wall or with sediments, which has been found to be significant. More research is therefore needed to understand the interactions on the pipe surface and with sediments.

With a growing number of water systems using chloramines to reduce THMs and HAAs, there is the risk of forming chloraminated based DBPs. The DBP that has received the most attention is N-nitrosodimethylamine (NDMA). It is being investigated by several researchers. However, a fundamental understanding of the formation of NDMA is needed before an effective model can be developed for the formation of DBPs following chloramination.

Microbial Water Quality

The results of numerous studies into the effect of different parameters on microbial water quality demonstrate that many interrelated factors play an important role in controlling microbial growth. The determining factor for deterioration of water quality with increasing retention time is whether the conditions within the network are conducive to the growth of the organisms of concern. High temperature, low disinfectant residual, high organic content and a greater surface area to which microbes can adhere will promote microbial growth. When these models are applied to distribution systems it is difficult to evaluate the extent to which the various processes occur and to define the interrelationships between them. Future research needs to:

- Assess the significance of alternative causes of high microbial counts and the implications for modeling microbial water quality
- Improve methods of characterizing the nature of NOM and how it controls biofilm and bacterial growth
- Research biofilm and its behavior and thereby develop methods for modeling biofilms in distribution networks

APPLYING MODELS TO DISTRIBUTION NETWORKS

Many chemical and microbial models simulate the behavior of an isolated vessel of water. For these models to be applied to real distribution networks, they need to be extended to take account of:

- Variation in pipe properties across a network, in particular pipe material, condition and diameter
- Variation in flow, velocity, and water age with time and space
- Mixing between water that has traveled along different flow paths

Water Quality Network Models

Distribution network models such as EPANET have been used to predict hydraulic behavior for many years. Hydraulic analysis is often predominantly concerned with peak flow conditions and the low pressures which result. However, for water quality modeling, the accurate prediction of travel time is fundamental. Hence, accurate definition of diurnal variations in flow covering both low and high flow periods is important.

Hydraulic network models have been integrated with models that simulate a range of different water quality parameters.

Chlorine and chloramine models: Models have been used by many water companies to simulate disinfection residuals. Studies have achieved good calibration between modeled and observed data. However, their long-term performance has been restricted by the use of over-simplified first order kinetic models.

DBP models: Proposed new US regulations (Stage 2 Disinfectants and Disinfectant By-Product Rule) specifically identify the use of water quality modeling. Utilities are likely to need guidance on the most effective modeling methods to comply with the regulations. For example, most existing models ignore pipe wall effects although there is some evidence that the pipe wall can influence DBP levels.

Discoloration models: Some network models aim to evaluate the risk of discoloration by comparing pipe flow predictions with predicted velocities for the deposition resuspension of a characteristic sediment particle. Their performance is questionable because they oversimplify and ignore many of the complex interactions which are necessary for discoloration.

Microbial models: Although models have already been developed for microbial behavior they are complex and utilize parameters that are not routinely monitored by utilities. Hence, they have not achieved widespread use.

Hydraulic Issues

There is a need to evaluate the importance of the various issues which govern the ability of hydraulic models to predict accurately the flow routing and travel times. Guidance is then needed on specific modeling methods and standards that should be achieved when using different water quality modeling applications. The guidance should cover:

- Model development, skeletonization, and effective use of auto-calibration
- Quality and quantity of base data
- The methods and accuracy of spatial and temporal demand allocation
- Calibration and validation
- Estimating model and predictive analysis accuracy and confidence limits
- Constraints associated with network design and operational management
- Measurement and modeling of tuberculation
- The application of models to conditions outside those from which they were derived or calibrated

Tracer studies offer the flexibility to measure travel time directly and to assess blending within a network. This information is valuable for calibrating or validating models. Although the principle behind tracer studies is straightforward, it is important to consider the appropriate trace chemical, duration of dose, and sampling method. High travel times and multiple flow paths can diminish the clarity of results. There is a need for guidance so that utilities can address these factors and make effective use of the technique

Water Quality Issues

There are a number of issues relating to the incorporation of water quality models within network models:

- Research on individual water quality parameters is continually improving individual process models. These developments need to be transferred into water quality network models, although some simplification of the model may be required to produce a tool which is practical for application by utilities.

- Network modelers often have an engineering background. However they need to understand the multiple relationships between physical, chemical and biological parameters and the impact of changes in one parameter on the others. This will allow them to recognize and understand the possible causes when models do not fit observed behavior.
- Temperature has a significant impact on most water quality parameters. Hence, modelers need methods for predicting temperature changes and accounting for their impact on model parameters.
- Water quality models can require numerous input variables and coefficients. A model can only be applied where it is possible to evaluate these variables. Guidance is needed on standard values for these coefficients and methods for measuring or estimating them.
- Water quality coefficients are sometimes treated by modelers as adjustable 'calibration' parameters. However, this approach can result in inaccurate values, especially if the model involves multiple coefficients.

Storage Tank Models

Tanks can be modeled either in isolation or as part of the distribution network. If modeled in isolation, it is important to consider how they interact with the distribution system and whether some of the connected pipe work is effectively part of the storage. The methods for modeling tanks include:

- *Physical models:* Scale models used to observe hydraulic behavior using dye or chemical tracers
- *Computational fluid dynamics:* The three dimensional modeling of fluids based on fundamental physical principles.
- *Empirical models:* There are three basic models for simulating fully mixed, plug flow or short circuiting behavior. More complex behavior within a single tank can be simulated by combining these models (known as 'multi-compartment' models).

Zone Level Models

Some modeling techniques are applied at a system level. Although they consider network properties, they do not have a point predictive capability beyond the zonal level. They can be used for comparing the relative performance of zones to identify area in which control measures are required. Zone level methods include:

- *Statistical models:* Statistical models are suitable for microbial parameters and can be applied directly to entire distribution networks. A limitation is the size of the data sets required to derive the statistical relationships of low frequency events.
- *Tabulated conceptual models:* These are useful where there is sufficient knowledge to enable a structured approach, but where detailed numerical modeling is impractical.

RESEARCH NEEDS

At the project workshop, eleven research briefs were proposed which were considered to address the most urgent needs of the industry. The titles of the proposed briefs are:

- Guidelines for Developing, Calibrating, and Using Hydraulic Models for Water Quality Modeling
- Fate and Characterization of Natural Organic Matter (NOM) in Distribution Systems to Develop Water Quality Models
- Susceptibility of Distribution Systems to Pressure Transients
- Reference Manual for Evaluating Water Quality Model Coefficients
- Impact of Pipe Wall Factors on Modeling Water Age, Chlorine, and Monochloramine Decay
- Internal Pipeline Scale Characterization
- Assessment of DBP Modeling Techniques for Compliance with the Stage 2 Disinfectant and Disinfectant By-Product Rule
- Characterization of Particles to Develop Models for Controlling Discoloration
- Decision Tool for Bacterial Regrowth Control Options
- Modeling the Health Risks of Microbial Intrusion into Distribution Systems Operated with Chlorine and Chloramine Residuals
- Tracer Studies for Water Quality Model Calibration and Validation

CHAPTER 1
INTRODUCTION

OBJECTIVES

The objectives of the project are to characterize the current state of predictive distribution system water quality models and identify critical research needs for their improvement. The scope of the project includes both the development and application of models. The report is intended to both steer future research and act as a general reference on water quality modeling.

The project was based around a workshop which evaluated the research needs and developed the research briefs. The workshop brought together delegates with expertise in microbiology, water chemistry, and hydraulic modeling. This enabled the cross-fertilization of ideas and the development of a complimentary research agenda. Hence, the report draws from the expertise and opinions of water company practitioners, academics, and consultants with a broad range of water quality, operational, and modeling expertise.

A major output from the work is a set of eleven research briefs which were developed at the workshop and have been submitted to Awwa Research Foundation (AwwaRF) for funding consideration (Appendix A).

WHAT ARE MODELS?

A model can be defined as a system which is used to understand the behavior of another system to which it is analogous in some way. Normally the model is a much simpler system than the system being modeled.

PURPOSE OF MODELING

The primary purposes of water quality models are:

- *Observational:* Some models just record and present the observed behavior of a system. They are most relevant when the behavior of the system does not change in time. Maps or Geographical Information Systems (GIS) are examples of this type of model.
- *Investigative:* Some models are built as a means of investigating the mechanisms which govern the system. If a hypothesis is proposed for the mechanism of a system, one way of testing it is to develop a model of the system which simulates the proposed mechanisms. If the overall model behaves in a similar manner to the system, this gives credence to the hypothesis.
- *Predictive:* These models are used to predict the behavior of a system temporally and spatially and the response of the system to changes in the environment in which it operates. It is this type of model which is the subject of this report. Predictive models can be applied in three different ways (or a combination of these):
  - *Spatial prediction:* Models can be used to predict the behavior of a system at a location, particularly where it is difficult or costly to measure directly the behavior. This application is relevant to the analysis of distribution networks where the operator must ensure that adequate quality is maintained in the entire network,

although it is impractical to collect regular samples from every location in the network to confirm that the required quality has been attained.

- *Time prediction:* Models can be used to predict the behavior of a system at another moment in time under a changed environment (either previous behavior, or more commonly future behavior). For example, chlorine residuals will vary throughout the year, primarily due to variations in demand patterns, temperature, and source water quality. If these changes in demand pattern and temperature can be forecast a model could be used to predict the accompanying changes in the chlorine residual.
- *Scenario prediction:* Models can be used to predict the behavior of a system when different conditions are imposed on it or if the structure of the system is altered. When models are used for design purposes, they are generally used in this way.

Predictive models are often extensions of observational or investigative models. Since modeling normally involves simplification and requires unconfirmed assumptions to be made about the system being studied, it is important that credence is given to their output by validating them against data about the observed behavior. The degree of validation required depends on the type of model (discussed below) and the number and universality of the assumptions within the model. Some distribution water quality models will be highly system specific and need validating on each network to which they are applied. For other models this may be less important.

TYPES OF MODEL

Most models are physical, conceptual, or numerical.

Physical Models

Physical models are physical systems that simulate the behavior of another system. Often they will be scaled down and simplified representations of the original system, but this is not always the case. These models may include chemical or biological reactions. Examples of physical models include:

- Scale models of storage tanks or other distribution facilities
- Jar tests for chlorine decay
- Pipe rigs or pipe loops

Physical models are subject to the constraints of scale effects. In particular the hydraulic behavior of water does not scale linearly with the physical dimensions. This can be accounted for, but large scale differences will reduce confidence in the output. When pipe rigs or laboratory techniques are used to simulate distribution networks, there is a need to carefully consider and control the inputs and boundary parameters, both of which are unlikely to be consistent in a real distribution network.

Conceptual Models

These are essentially descriptions of the behavior of a system and the assumed mechanisms which control it. While they can not be used to predict absolute values, they can be used to predict whether a change is likely to affect the system and what the potential consequences are. The model could take several different forms:

- Descriptive text
- Diagrams
- Flow charts

Numerical Models

Numerical models aim to define the behavior of a system by mathematical equations. For example, chlorine decay is often modeled as an exponential function of time. Simple numerical models can be solved using manual calculations, but computer programs are more commonly used. In addition to calculating the solution, computers can facilitate the setting up of the model and enhance the interpretation of results through graphical displays giving a complete modeling package. There are two distinct types of numerical models:

Empirical Models

These are derived from observing the behavior of the system and defining mathematical functions which reproduce the observed behavior. An example of this approach is when formation rate curves are produced for trihalomethanes (THM) or haloacetic acids (HAA) in which the observed experimental data are fitted to the most appropriate mathematical expression. These models take no account of the actual mechanisms which control the inter-reactions within the system. If the system shows a sudden step change in behavior for conditions just outside those from which it was derived the change will not be reproduced in the model. Hence, empirical models should only be used within the range of the conditions from which they were derived or have subsequently been validated.

Statistical and risk-based models: Models based on the probability of occurrence can be used in risk-based management decisions. These models are derived by identifying statistical relationships between datasets. These are in essence empirical models, but there is benefit in considering them separately. Rather than predicting an absolute value for a water quality parameter, these models will predict the likelihood that the parameter will fall within a specified range. This approach has been used with some success to model microbial water quality in distribution networks by analyzing data sets of regulatory water quality data.

Mechanistic Models

These models attempt to replicate the individual mechanisms which control the behavior of the overall system. This type of knowledge usually requires extensive research in a laboratory where the fundamental variables can be isolated and studied. An example of this is solubility modeling of pipe deposits such as lead or copper. Few models are fully mechanistic, at some level the component mechanisms are normally modeled via an empirical relationship. Mechanistic models are useful for developing an understanding of how a system works and in principle should be more robust when used to predict behavior under conditions for which the model has not been validated.

Empirical Versus Mechanistic Models

In practice most models are a combination of the above basic types. Thus, it is not uncommon to have an empirical model that is based on a partial mechanistic description of the system or a mechanistic model that has an empirical component. Empirical modeling tends to allow simplification of the model, however, it does not impart an understanding of why the system behaves as predicted, increasing the risk that it may be applied inappropriately. Mechanistic modeling gives a more holistic approach, but many water quality phenomena are too complex to define fully in mechanistic terms.

MODEL ACCURACY

Models often give absolute answers to a higher level of precision than is merited by the accuracy of the model. The modeler needs to have an understanding of the likely accuracy of models and structure their analysis accordingly. Models might be used to give knowledge on the expected behavior of a system even if the accuracy of the absolute answer is dubious. For example, a model may suggest that increasing pH will decrease chloramine decay; which may be a useful insight even if the predicted magnitude of the chloramine decay is inaccurate.

WATER QUALITY REACTIONS

A key issue for understanding and modeling water quality within distribution is whether the reaction or process occurs within the main body of the water (bulk reactions) or whether they occur on the internal surfaces of the distribution system (wall reactions). Many water quality parameters can be influenced by both processes.

Bulk Reactions or Processes

Bulk reactions are due to the interaction of the various chemical compounds or microbial organisms transported in the bulk flow of the water. Hence they are strongly influenced by the quality of the treated water entering the network. From the researcher's perspective, bulk reactions have the advantage that they are relatively straightforward to investigate and evaluate, because they can be observed by taking a sample of water and analyzing it in the laboratory.

Wall or Surface Reactions

Many water quality processes involve reactions with the pipe material itself or with material which has accumulated at the pipe wall. Hence, the condition and material of the pipe can have a significant influence on water quality. Of particular significance are unlined iron pipes which can cause rapid decay of disinfectant residuals, are prone to the release of iron causing discoloration, and promote biofilm growth.

Wall reactions are difficult to investigate due to the difficulty of observing behavior at the pipe wall. Furthermore pipe conditions often vary considerably throughout a network and so even when measurements are taken it can be uncertain whether they are representative of the entire network.

TYPICAL APPLICATIONS

Water quality issues affecting the industry are listed below, along with examples of how modeling can be used by utilities to help address each issue.

Disinfectant Residual

Most operators aim to maintain a minimum disinfectant residual throughout their system to control bacterial regrowth and protect against contamination without an excessive dose at source. In the USA the Surface Water Treatment Rule (SWTR) requires that 95% of the samples collected monthly must contain greater than 0.2 mg/L of disinfectant. The proposed disinfectant and disinfectant by-product rule (DBPR) has set new standards for maximum disinfectant residuals of 4 mg/L.

Possible Applications

- Investigating whether disinfectant residuals can be increased in part of a network by altering the operation and valve setup of the network
- Predicting the impact on distribution system residuals of increasing disinfectant dose, altering pH, or reducing natural organic matter (NOM) in the treated water

Disinfection By-Products

The Stage 2 DBPR will contain distribution based standards of 80 μg/L for THM and 60 μg/L for HAA. As a result more attention will be given toward controlling levels in DBPs in distribution systems. The regulations will also require utilities to demonstrate that representative sample locations are used:

Possible Applications

- Assessing the significance of NOM, disinfectant residuals, and retention time on levels of DBPs in a network
- Determining the optimal location of samples for regulatory monitoring based on the mean residence time

Corrosion of Metals

Lead and copper are common in domestic plumbing. Lead or copper can be released into drinking water and cause excessive levels. The degree to which this will occur depends on the chemical make up and properties of the source water. There is also concern with controlling discoloration, which results from the corrosion of unlined cast-iron pipes.

Possible Applications

- Use of solubility models to optimize treatment (pH, alkalinity, and orthophosphate) so as to minimize lead and copper solubility

- Investigating whether a discoloration problem might be controlled by altering water chemistry

Control of Nitrification

Chloramination is widely practiced in the USA as a means of controlling biofilms and minimizing the formation of DBPs. One of the problems associated with chloramination is that nitrification can eliminate disinfectant residual and potentially cause corrosion problems.

Possible Applications

- Investigating the scope for controlling nitrification in a network by altering the pH, chlorine to ammonia ratio, or reducing NOM

Control of Bacterial Regrowth

Excessive bacterial regrowth may cause violations of the total coliform rule (TCR) and contribute to corrosion and taste and odor problems.

Possible Applications

- Assessing to what extent disinfectant residual and organic content can be manipulated to control bacterial regrowth

Contamination or Intrusion of Pathogens

Distribution systems can be subjected to transient negative pressures due to abnormal pump operating cycles. These negative pressures may draw in water surrounding the pipe that is contaminated with pathogens.

Possible Applications

- Assessing whether areas of the network are at risk of negative pressure transients and whether theses regions overlap with regions with low predicted disinfectant residual

REPORT STRUCTURE

There are three separate parties involved in the development and application of water quality models each of which will have a different perspective on the research needs.

- *Chemistry specialists:* People with a scientific background who are involved in the research of the chemistry aspects of water quality such as DBP formation,
- *Microbial specialists:* Again people with a scientific background
- *Modeling practitioners:* People who aim to apply modeling developments to the management of distribution networks. Often these people have an engineering background with primary expertise in hydraulics rather than water quality.

This report starts by reviewing the current state of water quality modeling from each of these perspectives (Chapters 2 to 4). The conclusions from the report represent the conclusions from the workshop and are given in Chapter 5. Details about computational fluid dynamics (CFD) and artificial neural networks (ANN) are given in Appendix B.

CHAPTER 2
MODELING CHEMICAL AND AESTHETIC WATER QUALITY

This chapter covers modeling the processes of corrosion, discoloration, disinfection decay, and DBP formation. In general, the models and research presented in this Chapter either ignore the influence of the distribution network, or assume that the network has uniform properties throughout its entirety. Many of the models consider the influence of the pipe wall, but they do not account for the following:

- Variation in pipe properties within the network, in particular pipe material, condition, and diameter
- Variation in flow, velocity, and water age with time and space throughout the network
- Mixing between sources and water which has traveled along different flow paths

Chapter 4 reviews how water quality modeling can be extended to take account of these effects.

GENERALISED PROCESS MODELS

The following section presents some basic models which are frequently used to simulate water quality reactions. These models are generalizations and may not fully simulate complex interactions. Nevertheless the models are widely referred to throughout this report, often as components of more complex models.

Kinetic Models

The rate of most chemical reactions decreases as the concentration of the reactants decreases and can typically be approximated by an equation of the following form:

$$\frac{dC}{dt} = -k\{Reactants\} \tag{2.1}$$

where C = concentration of material
t = elapsed time
k = specific reaction rate constant (units dependent on order of reaction)
f$\{Reactants\}$ = some function of the concentration of the reactants

The reaction order is defined as the order of the differential rate in Equation 2.1. Where the reaction rate is directly proportional to the concentration of a single reactant, the reaction is classed as 'first order.' A 'second order' reaction can be due to the reaction rate being proportional to the square of the concentration of a single reactant, or due to the reaction rate being proportional to the product of two reactants.

First Order

$$\frac{dC}{dt} = -kC \tag{2.2}$$

or

$$C = C_0 \exp(-kt) \tag{2.3}$$

Second Order Single Compound

$$\frac{dC}{dt} = -kC^2 \tag{2.4}$$

or

$$C = \frac{C_0}{(1 + C_0 kt)} \tag{2.5}$$

Second Order Two Compounds

$$\frac{dC}{dt} = -kCH \tag{2.6}$$

or

$$C = \frac{H_0 S - C_0}{\frac{H_0 S}{C_0} \exp\left(\frac{k}{S}(H_0 S - C_0)t - 1\right)} \tag{2.7}$$

where C_0 = initial concentration
H = concentration of reactant
H_0 = initial concentration of reactant
S = stoichiometry of reaction

Many chemical reactions are complex, involving more than two different compounds and numerous intermediate steps. This means that the overall reaction kinetics will be complex and difficult to define. Hence, for pragmatic reasons it is often necessary to simplify the reaction kinetics. This can be justified where one part of the reaction is much slower than the rest so that it becomes the 'rate determining step.' Alternatively, where all but one of the reactants are present in large excess, and only the concentration of that reactant changes significantly during the course of the reaction, the reaction rate may appear to follow a first order pattern. This is sometimes referred to as a 'pseudo first order reaction.' For example, free chlorine decay is frequently modeled by a first order kinetic model (exponential decay with time).

Bulk and Wall Reactions

It is widely accepted that the concentration of many water quality parameters will vary due to reactions with compounds contained within the bulk water and due to reactions at the pipe wall. The overall reaction coefficient can be defined as the sum of a bulk and wall coefficient.

$$k = k_b + k_{w*} \tag{2.8}$$

where k_b = bulk water decay coefficient
k_{w*} = wall decay coefficient

It has been argued that the rate of consumption at the pipe wall will be controlled by the reactivity of the pipe surface and the rate at which the reactant can be transferred from the bulk flow to the pipe wall. There is little published research on detailed laboratory or field investigations into this process for water distribution specific conditions. A limited number of models have been developed which aim to simulate this process based on general theory of hydraulic mixing, chemical diffusion, and mass transfer. One model of particular note is the mass transfer model incorporated within the EPANET network analysis software (equation 2.9). In this model the rate of radial mass transfer is defined by the coefficient k_f, which is itself a function of turbulence and the molecular diffusivity (Rossman 1994, Rossman et al. 1994, Clark et al. 1995). General details about network analysis software and EPANET are given in Chapter 4. However, the mass transfer model is presented here because it details a specific process and some of the investigations into pipe wall effects presented in this Chapter are based on an acceptance of this model.

$$k_{w*} = \frac{4k_w k_f}{d(k_w + k_f)} \tag{2.9}$$

where $$k_f = Sh\frac{D}{d} \tag{2.10}$$

$$Sh = 0.23Re^{0.83}Sc^{0.333} \quad \text{for } Re \geq 2{,}300 \tag{2.11}$$

$$Sh = 3.65 + \frac{0.0668\frac{d}{L}ReSc}{1 + 0.04\left(\frac{d}{L}ReSc\right)^{0.666}} \quad \text{for } Re < 2{,}300 \tag{2.12}$$

$$Sc = \frac{\nu}{D} \tag{2.13}$$

$$Re = \frac{ud}{\nu} \tag{2.14}$$

k_f = mass transfer coefficient
k_w = wall reactivity coefficient
Sh = Sherwood number
Re = Reynolds number
Sc = Schmidt number
d = pipe diameter
L = pipe length
u = low velocity
D = molecular diffusivity
ν = kinematic viscosity

The above model has been widely used to predict free chlorine decay. The model is derived from theory for mass transfer in straight pipes used in the design of chemical process plants. There may be some limitations in applying this theory to distribution networks due to the effects of bends, non-uniform tuberculation, and the much greater pipe lengths involved, however, this has not been widely discussed in the literature. While many studies have validated the overall EPANET network modeling package, the studies have not generally been designed to test the mass transfer aspects of the model. Ideally, this would require careful measurement of wall decay as parameters in the model are varied. These parameters should include velocity, pipe length, pipe diameter and temperature (temperature controls molecular diffusivity and viscosity). Given the widespread use of the EPANET model, some form of validation against controlled tests on pipe rigs or isolated sections of a network would be prudent.

CORROSION AND METAL RELEASE

Large sections of distribution systems are composed of metals that are relatively unstable and subject to corrosion. The materials of primary concern for water quality are iron, lead, and copper. Corrosion can cause excessive levels of these metals within drinking water, a particular problem for lead and copper due to their toxicity. Iron corrosion can cause severe discoloration, production of sediment, and may promote microbial growth.

Over the past twenty years there has been a significant amount of research to develop “solubility models” for lead and copper. Solubility models can be used with a relative degree of accuracy to predict the concentration of soluble mineral compounds in drinking water. The models are valid only if the mineral compounds on the internal surface of the pipe are known. For most cases however, the composition of the compounds is assumed based on very limited data (Schock 1989). For lead and copper pipe work the crystalline mineral compounds have been identified using X-ray diffraction. The surface of iron pipes is a complex mix of amorphous (non-crystalline) and crystalline compounds (Clement et al. 2002) and, to date, iron solubility models have not been developed.

Solubility models can give firm guidance on the impact of water treatment adjustments on the concentrations of mineral compounds; for example changes in pH, dissolved inorganic carbon, and orthophosphate to minimize lead levels in distribution. A model for each mineral compound is derived using field and laboratory experiments to determine a solubility constant for the compound. This constant is then used together with other water chemistry data to predict how concentrations change with pH and dissolved inorganic carbon (Schock 1989).

Lead Release

Lead is a major concern due to its high toxicity, particularly for infants. Lead is regulated in most countries. In the USA the Lead and Copper Rule set an action level at 15 μg/L in 90% of samples. The European standards have set the limit at 10 μg/L, although how the standard is applied varies from country to country. Lead piping is still present in some service lines from the distribution main to a dwelling and in some internal plumbing.

One method to control lead concentrations in drinking water is to alter the water chemistry and minimize the solubility of lead. Solubility models for lead were developed by Schock (1989). The primary factors controlling lead in the solubility models are: pH, carbonate, and orthophosphate. The models are based on the premise that the lead solids are in equilibrium with the water. The assumed solids are hydrocerrusite ($Pb_2CO_3(OH)_2$), cerrusite $PbCO_3$, and various forms of lead orthophosphate. The limitations of these models are that the assumed solids must be present, which may not be the case. Also, it is possible that coatings of inorganic or organic material may exist on the surface limiting the relative accuracy of the models. Despite these issues solubility models for lead have been used successfully.

Copper Release

Copper can cause health problem for infants and is a major concern for receiving bodies of water such as streams. The copper limits for many receiving waters are lower than limits set for humans. Solubility models for copper have been used for decades. Recently they were expanded by Schock and Clement (1998) to account for orthophosphate, chloride, and sulfate. A major consideration in modeling copper is whether it is present as copper (I) (Cu^{+1}) or copper (II) (Cu^{+2}). The solubility of Cu^{+1} is several orders of magnitude lower than Cu^{+2}. The difficulty is that Cu^{+1} exists at very low redox potentials and it is often difficult to determine which solid is present. However, given the redox conditions of most water systems it is assumed that copper is mainly in the Cu^{+2} form.

Copper solubility is affected by pH, carbonate, orthophosphate, and sulfate. As pH increases over the range 6 to 9 significant decreases in copper solubility can be observed. Increasing carbonate increases the solubility of copper. Orthophosphate at 1 to 5 mg PO_4/L at pH 6.8 to 7.8 will result in a substantial decrease in copper concentrations.

Iron Release

The release of iron from an unlined cast-iron pipe can cause serious discoloration problems with consumers. The problem with modeling iron release is that the solid phase is composed of a mixture of known and unknown compounds (Clement et al. 2002). Much of the recent research work has focused on simple conceptual models that describe how iron is released from unlined cast-iron deposits.

The major aspects of iron release are dependent on several layers: a porous core, a shell-like layer, and a top surface layer (Sarin et al. 2000, Sarin et al. 2001). The release of iron depends on the porosity of the scale, whether the conditions are anoxic, and the presence of certain ions. Given the diversity of the nature of the scale and its complex reactions it would be difficult to develop a predictive model without more research. A further complication is that the various previous studies into corrosion have observed wide variation in the nature of corrosion deposits and the release mechanism (Marshall 2001). Due to this system specific behavior, any models derived should be treated with caution until they have been validated on a wide range of different systems and water types.

Pipe Loops or Pipe Rigs

Pipe loops or pipe rigs have been used for decades to predict changes in metal concentrations in distribution systems. The aim of most of these loop systems has been to predict changes in lead and copper levels for compliance with the Lead and Copper Rule. Recently pipe loops have also been constructed for predicting iron release or red water (Clement et al. 2002)

The basis for the design of these systems is to replicate pipe work and flow conditions within domestic plumbing or within distribution systems. AwwaRF has developed a standard design for a lead pipe rig, along with a standard operation and monitoring protocol. This system consists of 17 to 24 ft of lead pipe with alternating periods of flow and stagnation (Economic and Engineering Services Inc. 1990). During the early 1990s there were many pipe loops installed in the USA to predict lead and copper levels. Results were only published for a few of these studies, but the limited data suggest that the loops had very high relative accuracy and low absolute accuracy. Part of the weakness of this type of model is that usually it uses new pipes that are not representative of actual plumbing which may be over 50 years old. Even when aged piping material is used in a pipe loop it may still take up to 6 months before metal levels stabilize and the rig can be used for testing. Thus the total testing time may require 18 to 24 months.

There has been a decline in the use of pipe loops for predicting lead and copper since solubility models have proven to be very reliable with high relative accuracy. The area where pipe loops are having more use is for predicting iron concentrations or red water, as solubility reactions and the pipe scale components of iron pipe are poorly understood. Pipe loops for iron tend to be considerably larger than those used to investigate copper or lead solubility because they are used to replicate the behavior of distribution pipes rather than domestic plumbing. Construction and operation of these systems can be expensive as a length of pipe must be excavated from an existing network and transported to the test site. Both the Massachusetts Water Resource Authority (MWRA) and the Greater Cincinnati Water Works have successfully used iron pipe systems. The MWRA pipe loop system consisted of seven individual racks each consisting of 100 ft of 6 in unlined cast-iron pipe. The pipe loop system was also used to determine the effects of flushing and chlorine on general water quality.

DISCOLORATION

Discoloration of supplies is most commonly caused by fine suspended sediment rather than a true coloration of the water. There are two distinct components to modeling discoloration:

1. Modeling the source or the rate of production of the sediment
2. Modeling the entrainment, transportation, and deposition of sediment as it passes through the network.

Sources of Sediment

The following sources of sediment have been noted by Gauthier et al. (1996) and De Rosa and Miller (1994):

- Carry-over from the treatment works (comprising carry-over of coagulant from water treatment or inadequate removal of iron from the source water)
- Corrosion of metallic components, in particular unlined distribution mains and galvanized service pipes
- Precipitation of dissolved elements
- The growth of micro- and macro-organisms

Modeling Sources of Sediment

To date most attempts to model discoloration have assumed a constant sediment flux into the network. This might be reasonable for scenarios in which the rate of production is low and the discoloration is predominantly caused by the disturbance of sediment deposits which have accumulated over many years. However, when there is a more rapid production of sediment causing direct discoloration of supplies, the rate of production will be a limiting factor in the degree of discoloration and so needs to be modeled in more detail. In particular, further research is required to enable the numerical modeling of the rate of sediment production caused by:

- The 'bleeding' of iron mains (see *Corrosion and Metal Release* later in this chapter)
- Precipitation reactions when chemically incompatible waters are blended within a network
- Sediment caused by microbial growth

Sediment Transport

The movement of sediment in a water main may be considered in three broad modes, each of which needs to be accounted for in a comprehensive sediment transport model:

1. *Entrainment:* As the water flows through the water main, it imparts hydrodynamic forces on any particle resting within the pipe. If these forces exceed the gravitational and adhesive forces acting on the sediment, then the particle is dislodged and entrained in the flow.
2. *Transport:* This occurs when the flow velocity remains high enough to maintain the movement of the entrained material through the system. Finer and lighter materials tend to travel in suspension. Larger and denser particles roll or slide along the bottom of the pipe as 'bedload.' An intermediate mode of sediment transport is 'saltation,' in which the particles bounce along, but remain close to the bed. In suspended transport,

the particles effectively travel at the same velocity as the water, whereas in bedload transport it is substantially slower.

3. *Deposition:* If the flow velocity reduces, the sediment transport capacity of the flow reduces and material tends to be deposited along the pipe invert. The flow velocity at which all sediment movement ceases is generally less than the velocity required to entrain particles.

Once deposited, the sediment will gain some degree of adhesion to the pipe wall and adjacent particles due to cohesive forces. With time, chemical and biological processes may strengthen these bonds.

Characteristic Sediment Properties

Sediment transport is strongly controlled by the properties of the sediment, in particular the size, density, and concentration of sediment. For example, the settling velocity of a 100 μm size particle is approximately one hundred times greater than that of a 10 μm size particle. There has been relatively little research which characterizes sediment properties in water distribution networks and it is unknown how universal any published values would be. An additional complication is that the sediment properties may change as it passes through the network due to chemical reactions or the attachment of microorganisms.

Modeling Sediment Transport

Sediment transport has been widely researched over the previous 50 years and a number of different models have been developed. Most of this research was based on canals, rivers, estuaries, slurry transportation, or sewage and therefore relates to much higher sediment concentrations and larger particle sizes than is common in water distribution mains. Ackers, Brandt and Powell (2001) undertook a review of sediment modeling and recommended the use of several models which were derived from data which they considered to represent the closest to the conditions in water distribution mains. Much of the sediment which causes discoloration problems will be less than 20 μm in size. This is well below the minimum size on which any of the following models have been validated. Hence, the models are of use primarily for indicating the velocities at which sediment risks being disturbed or the minimum velocities for effective flushing rather than predicting the precise distribution of sediment throughout a network.

Macke (1982) Formula

This model predicts the velocity required to maintain a given transport capacity for a given sediment size. The sediment is assumed to be carried in suspension with no sediment bed. The method was calibrated on a minimum sediment size of 160 μm and a minimum sediment flux of 10 mg/L and so the results will be questionable if applied to significantly smaller sediment sizes or transport capacities. In particular, the method implies that the transport capacity reduces to zero only at zero velocity, which is clearly incorrect.

$$C_v = \frac{0.419\lambda^3 V_L^5}{(s-1)W^{1.5}d^2} \text{ (SI units only)} \tag{2.15}$$

where C_v = sediment transport capacity (as a volumetric concentration)
λ = Darcy-Weisbach friction factor
V_L = limiting velocity for deposition
W = fall velocity of sediment particles
d = pipe diameter
s = specific gravity of the solids particles

May (1994) Formulae

This method predicts the threshold velocity and the velocity required to maintain a given transport capacity for a given sediment size. The sediment is assumed to be carried as bed load. This method is probably more accurate than Macke for larger sediment (1 mm or greater). Unlike the Macke (1982) method, it can be used with transport capacities below 10 mg/L.

$$C_v = 0.0386\left(\frac{d_{p50}}{d}\right)^{0.6}\left(1-\frac{V_t}{V_L}\right)^4\left(\frac{V_L^2}{g(s-1)d}\right)^{1.5} \tag{2.16}$$

and

$$V_t = 0.125\sqrt{g(s-1)d_{p50}}\left(\frac{d}{d_{p50}}\right)^{0.47} \tag{2.17}$$

where d_{p50} = particle size which is larger than 50% by mass of all the particles in a graded sample
V_t = velocity at threshold of motion
g = acceleration due to gravity

Ackers and White (1973) Formulae

These formulae calculate the total concentration of sediment carried for a given flow and are applicable to fine and medium sand, from 60 μm up to approximately 0.6 mm. However, the method is only valid when there is a bed of sediment in the pipe. The low sediment concentrations in drinking water systems make it questionable as to whether this method is appropriate.

Adhesion

A major limitation of existing sediment transport methods is that they do not account for the adhesion of particles to the pipe wall and cohesion with other particles. A simple approach, as used in the modeling of sediment in sewers, is to consider a threshold shear stress (Ackers, Butler and May 1996). However, this cannot accurately simulate the complex combination of chemical, biological, and physical mechanisms which vary with time and which could also change the size and density of particles and the strength of adhesive bonds.

It is widely observed that, if the flow in a pipe is reversed, then sediment can be entrained even at very low velocities. This could be due to deposited sediment tending to form shapes which have lower drag in the direction of forward flow than the direction of reverse flow. It could also be due to sediment which has been trapped behind perturbations in the internal pipe surface, such as pipe joints or corrosion tubercles. This sediment would be held in place by hydraulic forces but, if the flow reverses, then this sediment can be released. The existing sediment transport methods do not account for this effect, which is a significant cause of discoloration.

DISINFECTANT DECAY

Due to the tightening of regulations controlling DBPs and the wish to reduce the taste and odor of chlorine, operators are under increasing pressure to reduce disinfectant residual exiting the treatment facility. However, there is still the need to protect against microbial contamination or regrowth and a disinfectant residual is required throughout the network as part of the SWTR. Modeling disinfectant residuals is an important tool for utilities as it allows them to identify areas at risk of low residual, predict how changes to the network will affect the residuals, and optimize the network set-up so as to minimize the required disinfectant dose.

This section covers modeling aspects that are used to predict levels of free chlorine and chloramine disinfectants. Both bulk and wall reactions are significant to disinfectant decay and so are considered separately.

Free Chlorine

Bulk Decay

Chlorine decay is frequently modeled as a first order reaction (Equation 2.2) and this has often given reasonable results. A variety of alternative models for bulk decay were tried in the AwwaRF study by Vasconcelos et al. (1996) into chlorine decay. Some of these gave better fits against observed data than the first order model. However, this was often a consequence of the models containing more variables and thereby greater mathematical degrees of freedom. These additional variables theoretically represent kinetic parameters and the concentration of other reactants, but, due to the difficulty in measuring these parameters, they were treated as calibration variables. Models with multiple variables are of limited benefit unless the variables can be related to measurable parameters. Similar tests by Powell et al. (2000b) on over 200 bulk decay tests also found models with multiple variables to give a better fit than the first order model. Nevertheless the first order model generally gave a good fit with a mean R^2 correlation coefficient of 0.95 (8 samples per test).

Some researchers have cited the following limitations with modeling chlorine decay as first order:

- It tends to underestimate the initial chlorine decay in the first four hours following chlorination. This can cause difficulties when fitting decay curves to jar test data because the value of the fitted decay coefficient will vary depending on:
  - The length of the delay between chlorination and the time of the first sample

  - The duration of the test
  - The temporal distribution of the samples
- It ignores the concentration of the principal reactants with which the chlorine reacts. If the concentration of these reactants in the source water changes significantly then the decay rate will change. In a network studied by Powell et al. (1999), the total organic carbon (TOC) concentration was observed to vary between 1.3 and 3.7 mg/L within a period of less than a month causing the estimated bulk decay coefficient to vary between 0.02 to 0.07 h^{-1} [0.5 to 1.7 d^{-1}]).
- The first order decay coefficient will decrease if the chlorine dose is increased and will also decrease significantly upon re-chlorination

Due to the first of these limitations, some researchers have separated the decay into two phases ('short-term' and 'long-term' demand).

To address the second two limitations, Vasconcelos et al. (1996) and Powell et al. (2000a) have had some success with using a first order model, but defining the decay coefficient as a function of the initial chlorine concentration, and some measure of the organic concentration of the water using a function similar to:

$$k = p\left(\frac{1}{C_0}\right)([TOC] - f) \tag{2.18}$$

where p = curve fitting parameter representing gradient of k versus $\frac{[TOC]}{C_0}$

f = curve fitting parameter representing offset of k versus $[TOC]$

All three of the limitations can be explained by the hypothesis that the various compounds with which the chlorine reacts limit the rate of the reaction. It is assumed that in the bulk water, NOM forms the principle reactants for chlorine. Unfortunately NOM usually consists of many different compounds. Some of these will be faster reacting than others and therefore depleted before other compounds (explaining limitation 1). A second order reaction (Equation 2.6) would allow the concentration of reactants to be taken into consideration. It has been used with some success to model the decay of chlorine in the laboratory when mixed with humic acid (Hua et al. 1999).

However, the second order model only allows for a single reactant. In practice the water will comprise a matrix of different chemicals with different reactivity and so will give a poorer fit for most real distribution waters. Another problem in applying the second order model is that it involves a second decay coefficient associated with the reactant. For the model to be viable for widespread application there is a need to derive some means of measuring this coefficient or of relating it to other measures such as TOC concentration. Clark and Sivaganesan (2002) have to a large extent addressed both these problems by simulating chlorine decay as two simultaneous second order reactions, one fast, the other slow, and by defining the kinetic parameters and the concentration of the reactants as functions of TOC, ultraviolet (UV) absorbance, initial chlorine concentration, initial bromide concentration, and pH.

Temperature Effects. The rate of virtually all chemical reactions increases with temperature. Bulk chlorine decay rates have been observed to increase with temperature by several groups (Zhang et al. 1992, Feben and Taras 1951, Jadas Hécart et al. 1992). The Arrhenius equation is widely accepted (Coulson and Richardson 1964) for modeling the effect of temperature on reaction kinetics:

$$k = F\exp\left(\frac{-E}{R(T+273)}\right) \tag{2.19}$$

where F = frequency factor (units dependent on order of reaction)
E = activation energy (J/mol)
R = ideal gas constant (8.31 J/mol°C)
T = temperature (°C)

An approximation by van't Hoff is that the reaction rates double for every 10°C rise in temperature (Fair and Geyer 1954). This is equivalent to a value of 5,750°C for E/R in the Arrhenius equation (Equation 2.19) for a temperature rise of 10 to 20°C.

In separate studies Vasconcelos et al. (1996) and Powell et al. (2000a) reported the Arrhenius equation to give a good fit against first order decay coefficients representing jar tests of free chlorine decay at various temperatures. This work covered a wide range of different water types. The optimum value for E/R ranged between 6,050°C and 7,300°C corresponding to a 2.1 to 2.5 fold rise in the decay constant between 10 and 20°C.

Wall Decay

As with bulk decay, decay at the pipe wall is also often modeled as a first order reaction. However, there is variation depending on whether the decay is defined in terms of a single wall decay coefficient (k_{w*} in Equation 2.8) or whether the wall reactivity (k_w) and the mass transfer rate (k_f) are modeled (Equation 2.9). In using a 'pseudo' first order model it is inherently assumed that the chlorine concentration is a rate limiting reactant. Vasconcelos et al. (1997) argued that although this may be the case for reactions with complex organic compounds such as those found in biofilms, it was less representative of the reactions with the corrosion products which line iron pipes. In such circumstances they stated that the chlorine would be immediately oxidized once in contact with the reductant and so the rate determining step is the rate at which the reductant is produced by the pipe. They proposed that a zero order reaction was more appropriate in these circumstances. In a zero order reaction the rate of chlorine consumption by the pipe wall is constant with time. They suggested that by using a zero order model, it might be possible to relate the reaction constant to corrosion rates.

Pipe walls can have a significant impact on disinfectant decay either due to reactions directly with the pipe material or due to reactions with material which has accumulated at the pipe wall. The pipe material strongly affects the rate of wall decay. Different pipe materials are listed below in approximate increasing order of their demand for free chlorine (Vasconcelos et al. 1996, NanCIE 1998, Powell et al. 1999):

- Polyethylene (k_{w*} negligible for new pipe)
- Poly vinyl chloride (PVC) (k_{w*} typically 0.05 to 0.5 h^{-1} [1.2 to 12 d^{-1}])

- Epoxy lined pipe
- Cement / bitumen lined pipe
- Unlined iron pipes (k_{w*} typically 1.0 to 10.0 h^{-1} [24 to 240 d^{-1}])

Unlined iron pipe has the strongest effect and where present is normally the limiting factor in disinfectant loss in distribution systems. Although it is possible to give a rough estimate on the expected order of magnitude for decay coefficients for different pipe materials, there can be considerable variability in the actual values. Valentine and Vikesland (2000) researched the role of the pipe water interface on disinfectant decay in unlined iron pipes. They found that in the systems they studied over 50% of the corrosion deposits were in the form of reduced iron and noted that in this state it will exert a significant chlorine or chloramine demand. This work along with research by Clement et al. (2002) demonstrates that a major problem in characterizing wall decay is that the deposit material can change significantly over a few meters and this will alter the chlorine or chloramine demand.

Velocity Effects. Some researchers (Woodward et al. 1996, Burgess et al. 1996, Powell 1998, Grayman et al. [N.d.]) have noted a tendency for increased rate of decay with increased velocity. The reason for this is not understood but it could be related to:

- The mechanism by which chlorine transfers from the bulk flow to the pipe wall. In particular, corrosion nodules could trap water which at low velocities could be virtually stagnant, allowing chlorine to be transferred across it to the pipe wall only by molecular diffusion.
- The disturbance of sediment which subsequently reacts with the chlorine

Chloramine

Like free chlorine, chloramine will progressively decay with time, but the rate of decay is typically much slower. Hence, chloramine is often modeled in a similar manner to free chlorine, but using lower decay coefficients. However, the decay process for chloramine is significantly different to that for free chlorine. Monochloramine is inherently unstable and will decompose producing free ammonia. This autodecomposition is strongly controlled by pH which is not accounted for by a simple first order model.

Bulk Decay

Some of the most extensive research into chloramine decay has been that undertaken by Valentine, Ozekin and Vikesland (1998). They showed that chloramine decay was controlled simultaneously by the multiple reactions involved in autodecomposition and reactions with NOM. They developed a mechanistic model that simulated fourteen interrelated chemical reactions. However, they also concluded that it was generally adequate to use a much simplified second order model (Equation 2.20).

$$\frac{1}{[NH_2Cl]} - \frac{1}{[NH_2Cl]_0} = k_{VCSC}t \qquad (2.20)$$

where $[NH_2Cl]$ = monochloramine concentration
$[NH_2Cl]_0$ = initial monochloramine concentration
k_{VCSC} = Valentine chloramine stability index

The value of k_{VCSC} varies with pH, initial chloramine concentration, inorganic carbon, and temperature. However, it can be calculated from measurable parameters and known rate constants and so does not need to be determined by experiment for each water.

Wall Decay

Several researchers have used iron pipe loops for predicting the amount of chloramine loss that may occur in distribution (Black & Veatch 2000). Although the loops showed consistent results, it was found that no major factor tested (dissolved oxygen, pH, alkalinity, and orthophosphate) affected chloramine decay. Also it appeared that the pipe loop was not useful in predicting in-situ chloramine loss.

Most modeling of chloramine wall decay has assumed first order kinetics. NanCIE (1998) performed a series of tests to compare the chloramine demand of various pipe materials with the free chlorine demand. They found that the relative reactivity of the different materials was the same with unlined iron pipes causing the most rapid decay and polyethylene causing the lowest decay. However, with all materials tested, chloramine was between four and ten times less reactive than free chlorine.

DISINFECTION BY-PRODUCTS

THMs and HAAs are formed by the reaction of chlorine or chloramine with NOM. The nature of the NOM and its reactivity with chlorine is highly diverse. As a result THM and HAAs are hard to model and tend to be based on empirical DBP formation rate data for specific water conditions. These empirical models often use TOC as a model input to represent all NOM, but this does not account for the reactivity of the specific mixture of NOM present in the water. Hence model coefficients can be highly site specific. For this reason it is advisable to perform bottle tests to measure the time formation curve for each water studied.

Most modeling of DBPs has focused on total THM formation due to reactions with free chlorine. Although some of these models have also been applied to HAA formation, only one model reviewed in the following section was specifically developed for HAAs (Westerhoff et al. 2002, mechanistic model). This model simulates THM and HAA formation within treatment plants.

THM formation reactions can occur rapidly and may be substantially complete prior to entering the distribution system, in which case they may be treated as a conservative substance. More commonly there will be some continued reaction within distribution with the result that THM growth within distribution can be as significant as their growth within the treatment plant. Clark et al. (1994) demonstrated that THM growth tended to decrease with time asymptotic to a maximum formation potential. This can be modeled using a first order saturation growth model (Equation 2.21).

$$\frac{dC}{dt} = k_b(C_L - C) \tag{2.21}$$

where C_L = limiting concentration

The limitation of this model is that it does not take account of the concentration of the various pre-cursors which control THM formation. Hence, the growth coefficient will be highly site specific and if the water is not of a consistent quality the coefficient may show considerable variability over time. To some extent this limitation is overcome by the method proposed by Clark and Sivaganesan (1998). This simulates chlorine decay and the formation of by-products as a second order model and assumes that the by-products represent total THMs. Expressions were developed for estimating some of the parameters in the second order model using TOC, pH, and temperature measurements.

More detailed models have been developed for the formation of THMs within the water treatment process and in particular during super-chlorination. Amy et al. (1987) proposed an empirical model of the following form which has subsequently been incorporated into the USEPA water treatment plant (WTP) model (Equation 2.22):

$$[THM] = X_0[\mathrm{TOC}]^{X_1} pH^{X_2} t^{X_3} T^{X_4} [Br]^{X_5} [Cl_2]^{X_6} [UV]^{X_7} \tag{2.22}$$

where $[THM]$ = THM concentration at time t
X_0 to X_7 = WTP fitted coefficients
$[TOC]$ = TOC concentration
T = temperature (°C)
$[Br]$ = bromine concentration
$[Cl_2]$ = chlorine dose
$[UV]$ = UV absorbance

The model was derived from data collected from the treatment process rather than distribution where there were higher chlorine doses and shorter contact times and hence their validity to the conditions within distribution is uncertain. Nevertheless, models of this type have been applied to distribution with some success for both THMs and HAAs (Seidel et al. 2002). A difficulty in applying this model to distribution networks is that the concentration of the variables may not be known. In addition, if the model coefficients (X_0 to X_7) are derived from jar tests, they will not account for interactions with the pipe wall or sediment which Valentine and Vikesland (2000) found could be significant. Table 2.1 lists the values for the coefficients to the above model, fitted by Westerhoff et al. (2002) to the formation of THM_4 and HAA_9, based on a series of approximately 100 chlorination experiments.

Westerhoff et al. (2002) has developed a detailed mechanistic model of DBP formation (THMs and HAAs). The model simulates reactions between chlorine, bromine, and NOM, with the NOM subdivided into instantaneous, fast, and slow reacting components. The model incorporates over 70 terms. Many of the constant terms have been evaluated through experimental work and calibration against data from 18 different treatment plants. This leaves a reduced number of site specific or variable terms to be evaluated when applying the model. The model was found to

Table 2.1
Fitted values for coefficients to THM / HAA formation model

	Coefficients to equation 2.22*							
Parameter / water source	X_0	X_1	X_2	X_3	X_4	X_5	X_6	X_7
THM_4								
Source 1	11.0	-0.101	1.20	0.250	0.408	0.367	0.309	0.689
Source 2	28.8	-0.518	1.40	0.281	0.191	0.270	0.325	0.827
Source 3	0.997	0.536	1.23	0.242	0.201	0.0365	0.615	0.0031
Source 4†	27.9	-0.328	1.32	0.257	0.197	0.122	0.261	0.858
HAA_9								
Source 1	231	-0.162	0.124	0.218	0.117	0.0318	0.323	1.14
Source 2	2.73E7	-1.43	-0.961	0.187	-0.301	0.296	-1.73	2.57
Source 3	1.98	1.39	-0.438	0.220	0.0239	0.192	0.438	-0.626
Source 4†	4.04E5	-0.205	-2.66	0.132	-0.260	0.593	-0.758	0.512

Source: Westerhoff et al. (2002)

* The units of the coefficients vary with their value. They are consistent with the use of the following units in equation 2.17:

$[THM]$ (μg/L)
$[TOC]$ (mg/L)
T (°C)
$[Br]$ (mg/L)
$[Cl]$ (mg/L)
$[UV]$ (cm^{-1})

† Source 4 is a composite of sources 1, 2 and 3

be more accurate than the empirical model developed by Amy et al. (1987) (Equation 2.22) when applied to source specific waters, but less accurate when applied to variable water sources. The model is intended for application to treatment plants and the authors are unaware of its application to distribution networks.

Recent research (Valentine and Vikesland 2000) suggests that the pipe wall effect on THMs and HAAs can be significant. In some cases organics bound to iron pipe deposits can cause an increase in DBPs, while adsorption onto iron pipe deposits can cause a decrease in concentrations. The preceding models do not account for these pipe wall effects which can be particularly important for systems with a significant amount of iron pipe. In such systems predicting DBPs could be difficult.

With a growing number of water systems using chloramines to reduce THMs and HAAs, there is the risk of forming chloraminated based DBPs. The DBP that has received the most attention is N-nitrosodimethylamine (NDMA). The formation of NDMA is being investigated by several researchers. However, a fundamental understanding of the formation of NDMA is needed before an effective model can be developed.

SUMMARY AND RESEARCH NEEDS

Generalized Process Models

- Kinetic models are widely used to generalize water quality reactions. First and second order models are a common component of many of the more complex models presented in this report.
- Some models simulate the radial mass transfer of constituents from the bulk water to the pipe wall. These are based on theory of mass transfer in chemical process plants. There is a need for detailed investigation of the mass transfer mechanisms in conditions directly applicable to water distribution to evaluate the significance of mass transfer processes and validate these models.

Corrosion and Metal Release

- Use of solubility models for predicting lead and copper levels has been found to be highly successful
- The use of pipe loop systems for predicting lead and copper levels, while relatively accurate, has been widely superseded by solubility models, which have a comparable level of accuracy at lower cost
- Pipe loops for predicting water quality effects on iron concentrations or red water have been shown to be useful in the absence of solubility models
- Research should be conducted to understand the compounds that reside on the internal surface of corroded iron pipes. With this information solubility models could be further developed to help predict treatment changes for controlling discoloration.

Discoloration

- In order to model discoloration it is necessary to consider both the rate of sediment production and the transportation of the sediment through the network
- Sediment transport models are predominantly derived from research on rivers, estuaries, sludge, or sewers. The validity of such models to water distribution, where the sediment is smaller and the concentrations lower, is questionable.
- The ability to accurately model discoloration is inhibited by several factors. Research is needed to redress these limitations:
  - There is a lack of numerical models for the rate of production of sediment, especially when it is caused by corrosion or microbial activity
  - No models available which fully account for the adhesion and erosion of sediment to pipe walls
  - Typical sediment properties are poorly characterized

Disinfectant Decay

- There are limitations in using a first order model to predict bulk free chlorine decay predominantly arising from its failure to account for the concentration of reactants. Second order models offer the potential to address these limitations, but involve additional model input variables. Methods have been proposed for estimating these variables from measurable parameters, but there is a need to apply these methods to the modeling of full distribution networks to test their viability for more general use.

- Temperature has a strong influence on the decay rate of free chlorine, but can be accounted for by using the Arrhenius equation (Equation 2.19)
- Chloramine decay involves different processes to those of free chlorine decay and so different models are likely to be required. In particular a first order decay model is less likely to be suitable for chloramine than it is for free chlorine. Alternative models for the bulk decay of chloramine have been developed which need to be applied to the modeling of full distribution networks to test their viability.
- There is a need for increased understanding of free chlorine and chloramine loss due to the pipe wall effects, with the aim of predicting how this will vary within and between distribution networks with apparently similar pipe work

Disinfection By-Products

- THM and HAA formation can be inconsistent in systems with a significant amount of iron pipe. DBPs may adsorb to the pipe and/or be degraded. More research is needed to understand the interactions on the pipe surface.
- Many existing DBP models are empirical and highly site specific. This is thought to be largely due to variations in the reactivity of the NOM mixture in the water. Mechanistic models would enhance the understanding of how different factors control DBP formation, but they require improved methods for NOM characterization.
- There is a need to develop models for the formation of DBPs following chloramination.

CHAPTER 3
MODELING MICROBIAL WATER QUALITY

This chapter reviews the modeling of microbial processes which occur within distribution networks. As with the preceding chapter, the application of these models to modeling entire distribution networks is considered in Chapter 4.

FACTORS WHICH CONTROL MICROBIAL WATER QUALITY

What Is Microbial Water Quality?

Water entering distribution networks is not sterile but contains a natural mixture of bacteria which is unlikely to adversely impact public health. Treatment of water ensures the absence of pathogenic microorganisms, primarily those that cause intestinal illness. Monitoring for the presence of specific pathogens in drinking water is impractical. Instead, evidence of fecal pollution from man or animals is determined by monitoring for the presence of indicator organisms. There are a number of different tests for microbial water quality:

- *Escherichia coli* is a natural inhabitant of the intestine of man and other warm blooded animals and birds. If present in water it is likely to indicate fecal pollution.
- Coliform comprise many different species and are found widely in the environment. Their presence in water does not necessarily indicate fecal pollution. Research has shown that environmental coliforms can grow in distribution systems.
- Heterotrophic plate count (HPC) or standard plate count is a count of viable bacteria on selected agar media. It provides useful information on changes water quality.

Parameters That Control Microbial Water Quality

There have been many studies that have determined the effect of different parameters on microbial water quality within distribution systems. The results of these studies demonstrate that many interrelated factors play an important role in controlling microbial water quality, including:

- Measures of carbon or NOM
- Temperature
- Retention time
- Dissolved oxygen
- Disinfectant residual
- pH
- Pipe material and condition

The balance between NOM and disinfectant residual is of particular interest since NOM promotes the growth of microbes whereas the disinfectant kills them.

Many studies have reported a correlation between coliform or heterotrophic bacterial counts and lack of disinfectant residual. Since retention time has a direct control on disinfectant residual, it is often assumed to control microbial activity indirectly. However, this is only valid where the disinfectant residual decays to a concentration which allows microbiological growth.

The determining factor for deterioration of water quality with increasing retention time is whether the conditions within the network are conducive to the growth of the organisms of concern. High temperature, low disinfectant residual, high organic content, and a greater surface area to which microbes can adhere will promote microbial growth.

Mechanisms

The mechanisms by which microbes grow or are killed within distribution are highly complex and probably vary between systems. The following generalized mechanisms have been cited:

- *Breakthrough:* The treatment process does not produce sterile water and so some microorganisms will pass through into the network from the raw water
- *Propagation of substrate, bulk organisms, and disinfectants:* This is principally governed by the hydraulic behavior
- *Growth:* This is the increase in bacterial numbers due to cell reproduction. Growth normally requires a source of carbon. It is not only the quantity of carbon which is important but also the speciation of the carbon. Laboratory tests have shown that the growth rate for biofilms exposed to an amino acid or carbohydrate solution can be six times higher than if grown in a humic acid solution (Camper 2003). An understanding of the nutritional conditions within the pipe network, therefore, is important to the modeling of growth within the distribution system.
- *Death:* This includes killing by disinfectants, lack of nutrients, and predation by competing microorganisms
- *Attachment and detachment:* The process by which microbes pass between the bulk flow and the biofilm. This process will be influenced by the hydraulics and the condition of the biofilm. The process may be periodic and localized with detachment occurring in clumps. As a result this process is difficult to evaluate.
- *Chemical processes:* Processes such as disinfectant decay and NOM utilization need to be considered due to the influence of these parameters on microbial growth or death
- *Surface processes:* The nature of the internal surface influences the susceptibility to biofilm attachment and growth. Hence, processes such as corrosion and scale formation may be important. Biofilm density has been shown to be dependant on pipe material (Camper 2003) with the density decreasing in the following order: unlined cast or ductile iron (highest biofilm density), cement lined iron, epoxy lining, and PVC (lowest biofilm density).
- *Absorption:* The process by which substrate and disinfectants are absorbed into biofilms and microbes

Clearly it is very difficult to model all the above processes and it may be unrealistic to expect highly accurate predictive models.

DISINFECTION MODELS

Models of the disinfection process at treatment plants have been used for many years. They are of interest because disinfection is one of the main processes which have to be accounted for in a model for distribution microbial water quality. Their applicability to distribution networks should be treated with caution due to the much shorter retention times and higher disinfectant doses involved.

The earliest models for the inactivation of microorganisms by disinfectants were the Chick and Watson models proposed in 1908 (Equation 3.1) and a generalization made by Hom in 1972 (Equation 3.2) (cited in Haas and Joffe 1994):

$$\ln\left(\frac{N}{N_0}\right) = -k^* C^{n'} N \tag{3.1}$$

$$\ln\left(\frac{N}{N_0}\right) = -k^* m C^{n'} t^{m-1} N \tag{3.2}$$

where N_0 = number of microorganisms at time zero (dimensionless)
N = number of surviving microorganisms at time t (dimensionless)
C = concentration of disinfectant (units dependent on disinfectant)
n' = coefficient of dilution (variable units)
m = Hom disinfection kinetic parameter (variable units)
k^* = pseudo first order reaction rate constant

More recently Haas et al. (1996) combined the Chick Watson and Hom models with a first order decay model to produce:

Modified Chick-Watson model:

$$\ln\left(\frac{N}{N_0}\right) = \frac{-k^* C_0^{\,n'}}{n'k^*}[1 - \exp(-n'k^* t)] \tag{3.3}$$

Modified Hom model:

$$\ln\left(\frac{N}{N_0}\right) = -\left(\frac{m}{n'k^*}\right)^m k^* C_0^{\,n'}\left[1 - \exp\left(\frac{-n'k^* t}{m}\right)\right]^m \tag{3.4}$$

Longer contact time and higher disinfection concentrations cause greater inactivation of microorganisms on contact with chlorine. The Chick Watson model produces an adequate fit of inactivation by free and combined chlorine under demand-free conditions according to results of Haas and Karra (1984), however Pernitsky et al. (1995) stated that the Hom model was found to be better in all disinfection streams studied. Ransome et al. (1995) found that the rate of inactivation is not linear and that the response showed an initial rapid kill which diminished after a 90% reduction in bacterial numbers had been achieved.

BACTERIAL GROWTH MODELS

A number of models are used by biochemical and waste water engineers to simulate the growth of microorganisms in unit processes such as activated sludge treatment of waste water (Metcalf & Eddy 1991). These models often involve much higher concentrations of substrate and microbes than would occur in distribution systems and generally apply to well defined systems for which it is relatively straightforward to define the model boundary conditions. The models will not describe the complex interactions which occur within distribution and so are not described in detail herein. Nevertheless, they demonstrate that it is feasible to model microbial processes with reasonable accuracy providing they are well understood. Some models such as the Monod equation (3.5) may be of some relevance for modeling the growth process.

$$\mu' = \mu'_m \frac{S'}{K_S + S'} \tag{3.5}$$

where μ' = specific growth rate (time^{-1})
μ'_m = maximum specific growth rate (time^{-1})
S' = concentration of growth limiting substrate in solution (mass / unit volume)
K_S = substrate concentration at half the maximum growth rate (mass / unit volume)

BIOFILM MODELS

There has been considerable interest in the occurrence of biofilms within distribution systems and the risk that they may promote microbial regrowth. Biofilms are organic deposits consisting of microorganisms, microbial products, and detritus (Marshall 1976, Characklis 1981, Safe Drinking Water Committee 1982). Biofilms may occur on pipe surfaces, sediments, corrosion tubercles, suspended particles, or virtually any substratum immersed in the aquatic environment. Biofilms may be evenly distributed or occur as sporadic random patches on a surface.

Biofilm formation involves: the transport and accumulation of microorganisms and nutrients at an interface; followed by metabolism, growth, and product formation; and finally detachment, erosion, or sloughing of the biofilm from the surface. The rate of biofilm formation depends on the physicochemical (chemical, thermodynamic) properties of the interface, the physical roughness of the surface, and physiological factors of the attached microorganism. Sheer forces generated by fluid velocity and possible effects of disinfectants on extracellular polymer substances may be important in the release of biofilms from surfaces.

Gatel et al. (1995) Empirical Model

Gatel et al. (1995) developed an empirical model to predict the bacteriological water quality in a given week if the water quality of the previous week is known. The model was based on data collected in the suburbs of Paris, 1992-1994. Results concluded that the model was strongly influenced by the general biomass quantity as measured by epifluorescence. Temperature and chlorine appear to be less important factors, however, this partly reflects the events chosen to develop the model. In addition, the biofilm may have been acting to protect bacteria from disinfection.

Lu, Biswas and Clark (1995) Mechanistic Model

A mechanistic model was produced by Lu, Biswas and Clark (1995) incorporating a range of interactions within a straight pipe. This two dimensional transport model is highly complex, accounting for the simultaneous transport of substrates, disinfectants, and microorganisms. The resulting model involves over forty different coefficients. It has not been verified by experimental data, but is useful for conceptualizing the nature of the relationships between the various mechanisms likely to be involved. The authors concluded that three significant factors were important in influencing water quality.

1. Shear stress determines the rate at which biofilm sloughs from the pipe surface
2. Radial diffusion in turbulent flow may either promote biofilm growth by providing nutrients at the pipe surface or inhibit growth by transferring disinfectant. The relative magnitude of both nutrients and disinfectants is therefore significant.
3. Pipe materials with a high chlorine consumption rate will cause low chlorine concentrations at the pipe-water interface and reduce control of biomass accumulation

SANCHO Model

The SANCHO model (Servais et al. 1995) predicts fixed and free biomass as measured by epifluorescence microscopy against residence time. It simulates a distribution system as a series of continually stirred tank reactors (CSTR) each CSTR representing a different finite section of pipe. Interactions between various parameters are modeled as detailed in Table 3.1.

The model breaks down the process into a number of mechanisms which are simulated by empirical relationships. Thus it is a semi-mechanistic model. Biodegradable organic carbon (BDOC) is assumed to be the limiting nutrient for microbiological growth.

SANCHO has been applied by Laurent et al. (1997) to three distribution systems. The surveys gave a reasonable correlation between modeled and observed concentration of free biomass ($R^2 = 0.69$). In the context of microbial modeling this is good performance. However, limitations with some aspects of the approach were noted by Gagnon et al. (1997).

Servais et al. (1995) further developed the SANCHO model to predict the changes in chlorine concentration, BDOC, and the biomass of free and attached bacteria in pipes where the diameter decreases as a function of residence time. The variables in the model include free-floating bacteria, adsorbed bacteria, three different fractions of BDOC, and free chlorine. Reaction kinetics and numerical values for the variables were determined from laboratory experiments.

Unified Biofilm Approach

Van der Kooij et al. (2002) developed the "Unified Biofilm Approach" for assessing the biological stability of water to enable the appropriate design of water treatment works and distribution systems to control regrowth, especially when disinfectants are not used. In the Netherlands, where

Table 3.1
Interactions modeled by SANCHO

Parameters	Change in concentration modeled by relationship with
Rapid reacting BDOC	• Degradation by enzymes • Increase due to cell death
Slow reacting BDOC	• Degradation by enzymes • Increase due to cell death
Substrate	• Formation from BDOC by enzymes • Utilization by bacteria
Fixed bacteria	• Growth due to substrate availability and inhibited by chlorine • Limited by space available for growth • Cell death
Adsorbed bacteria	• Rate of adsorption of free bacteria • Rate of desorption • Cell death
Free bacteria	• Growth due to substrate availability and inhibited by chlorine • Loss by adsorption • Gain from desorption
Chlorine	• First order decay

Based on: Gagnon et al. 1997.

water is distributed with either low or zero disinfectant residual, the Unified Approach allows the operator to predict any regrowth potential. The approach is based on three observations:

1. Assessment of the biofilm formation rate of water leaving the treatment facility in addition to determining the assimilable organic carbon (AOC) concentration. This is determined by allowing biofilms to develop on glass rings in a biofilm monitor and then measuring growth.
2. Assessment of the biofilm formation potential of synthetic materials in contact with drinking water. This is determined by measuring the concentration of active biomass on pieces of pipe material incubated in slow sand filtrate at 25°C.
3. Assessment of the biofilm concentration on pipe walls. This is determined from observations of biofilm growth on pipes at the treatment works or in distribution.

The amount of biofilm growth derived from the three tests is measured using adenosine triphosphate (ATP) to determine biomass and compared directly. These observations allow a ranking of waters and materials and together they provide a framework for predicting biostability in the distribution system.

HPC AND COLIFORM MODELS

Water utilities are primarily interested in modeling the parameters which are used on an operational basis to monitor water quality and demonstrate compliance with regulations. Predominantly these are HPC, coliform, and *Escherichia coli*. A major difficulty with modeling these is that in well-run systems, the levels of these parameters are extremely low and the flux of microorganisms may be unsteady or intermittent. This makes it difficult to collect sufficient data from which to derive or calibrate models. It is also possible that a significant proportion of the observed levels are due to factors other than microbial growth such as contamination of samples or intrusion. This makes it difficult to define the processes involved and develop mechanistic models. However, a number of attempts have been made using statistical models to predict the likelihood of microbial failures. Where sufficient data is available, it is possible to demonstrate statistical associations between the factors that affect water quality in distribution systems.

Statistical Models

LeChevallier et al. (1991) demonstrated through statistical analysis that coliforms were associated with a number of factors including nutrients (as measured by AOC), temperature, disinfectant type, and disinfectant residual concentration. They also found that a model produced by multiple linear regression based on free chlorine residuals and temperature predicted 83.8% of the variation in HPC bacterial numbers. The model predicted that at 20°C, a free chlorine residual of 1 mg/L would be necessary to maintain HPC below 500 cfu/mL.

Volk and Joret (1994) developed a simple and logical model to identify circumstances where there is an appreciable risk that coliforms may occur in drinking water. The model incorporates four water quality parameters: temperature, bacterial density (as measured by epifluorescence), free chlorine residual, and BDOC. For each of these variables, graphical analysis of the number of coliforms in samples against the respective measurement of the water quality parameter was used to derive a 'threshold' defining the range in which coliforms can be present. The four thresholds then provide a basis for scoring each sample on a five point scale, according to whether 0, 1, 2, 3 or 4 of the threshold values have been violated. This score gives an assessment of risk, which the authors link informally to the observed occurrences of coliform-positive samples. Depending on this score, sampling sites are described as 'without risk' (0 or 1 thresholds violated), 'weakly exposed' (2 or 3 thresholds violated), or 'exposed' (4 threshold values violated). Gatel et al. (1995) used a similar model for predicting coliform occurrences.

Gale et al. (1997) used a statistical approach based on the concept of "generalized linear modeling" (GLM) to develop coliform predictive models. Their aims were to:

- Identify statistical associations between parameters measured as part of the UK statutory water quality monitoring program
- Develop and validate models for predicting coliform occurrences
- Demonstrate the use of the models in maintaining and improving water quality

The results of the study show that coliforms were associated with higher numbers of heterotrophic bacteria, temperatures, and lower total chlorine residuals. Associations were also shown with coliforms, color, and turbidity. Models were calibrated by GLM to predict the probability of detecting coliforms in statutory samples from specific zones within UK water companies. Validation exercises indicated that the models were suitable for cautious application within the zones in which they were calibrated and in related zones within the same company.

The models can be used for predicting coliform occurrences after changing the associated water quality parameters. Figure 3.1 shows the relation between coliform occurrences and temperature. Based on these plots a UK operator has made an informed decision to take remedial

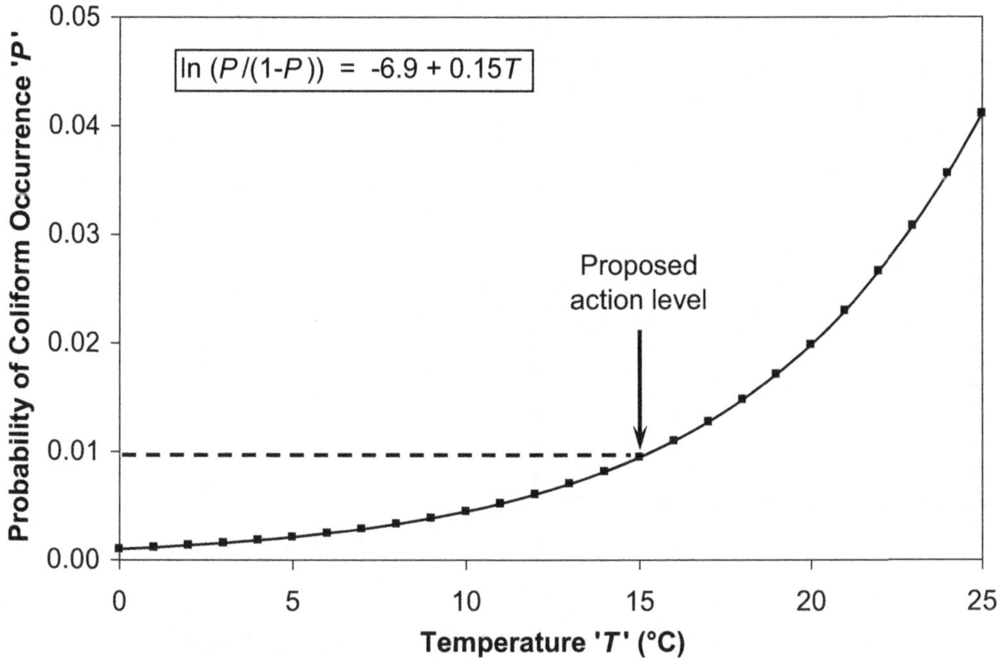

Source: Adapted from UKWIR report 97/DW/02/3 (WRc 1997)

Figure 3.1 Coliform predictive model 1

action when temperatures exceed 15°C such as deliberately drawing down storage levels to turnover stored water. In Figure 3.2 the relationship between coliforms and chlorine is shown. In this case the operator maintains chlorine residuals above 0.2 mg/L throughout the system.

A limitation of this approach is that all data is aggregated to zonal statistics before looking for correlations. Individual extremes in any of the data may be directly attributable to extremes in related water quality parameters for a given sample. For example, a coliform failure may be directly related to the high temperature of a sample. However, by aggregating the data together, the significance of these extremes in the final model is diminished.

The success of statistical modeling is dependent upon the quality and quantity of the data analyzed. The small number of coliform positive samples (<1%) used in Gale's (Gale et al. 1997) study make the models highly susceptible to both statistical and sampling errors. However, statistically significant results were produced with a positive influence on coliform failures produced by increases in temperature and HPC and decreases in total chlorine.

Further models for predicting microbial water quality or the risk of microbial failure are currently under development by other parties, but there are few published details to date. One of the goals of researchers has been to produce a generic model that can be applied to all distribution systems. In many cases these have been based on detailed research on one particular system. However, during the 1990s a number of researchers created tools or "rules of thumb" from reviewing literature and expert opinion in an attempt to derive generic models that can be applied to many, if not all, distribution networks.

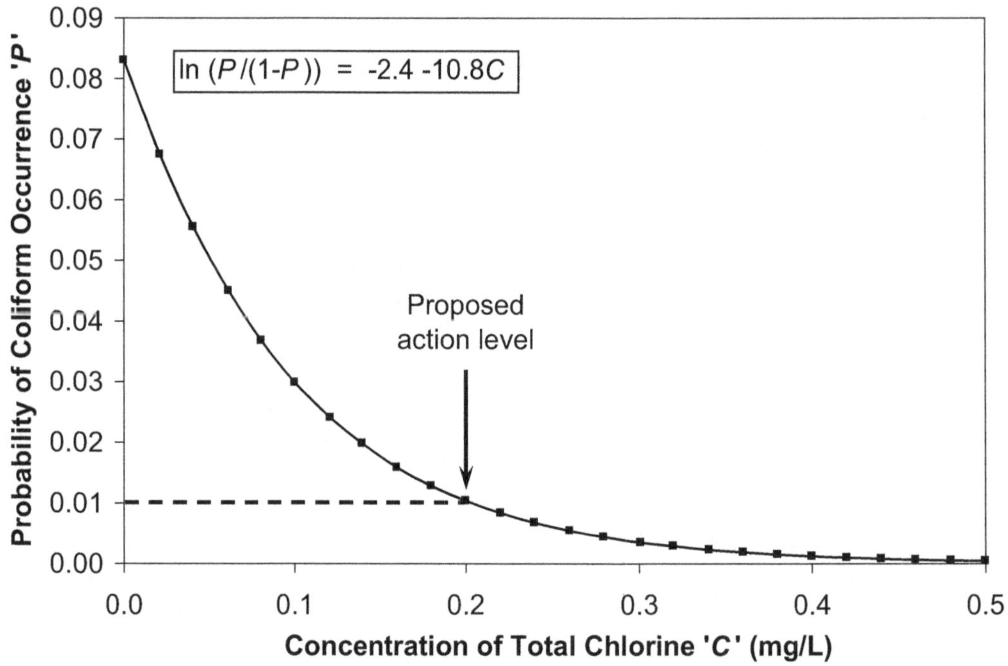

Source: Adapted from UKWIR report 97/DW/02/3 (WRc 1997)

Figure 3.2 Coliform predictive model 2

SUMMARY AND RESEARCH NEEDS

There is a reasonable understanding of many of the processes which control microbial water quality in distribution systems and some attempts have been made to model them. Some models for individual processes such as growth or disinfection have been used for many years. However, when applied to distribution systems it is difficult to evaluate the extent to which the various processes occur and define the interrelationships between them. Specific knowledge gaps and research needs are listed below:

- Some recorded microbial counts may be due to factors other than microbial regrowth such as intrusion and contamination of samples. These alternative causes may be highly erratic. It is important to assess the significance of alternative causes of high microbial counts and the implications for modeling microbial water quality.
- Many of the existing models may be highly system specific and possibly characterize existing performance (an observational model) rather than have true predictive capability.
- Various forms of NOM will provide the principle substrates for microbial growth. There is a need for improved methods of characterizing the nature of NOM and how it controls biofilm and bacterial growth, along with simpler techniques for its measurement.
- Compared with chemical parameters, there is often much less data available which can be used to develop models. Sampling methods are more laborious and on-line sensors

are not available for the main parameters of interest. Modeling coliform in particular is restricted by the very low counts which occur in well-run distribution systems.

- The occurrence of microbes in the bulk water due to detachment from biofilms is a heterogeneous process, which results in an unsteady flux of microbes. This random element is difficult to model explicitly.
- Extensive research into biofilms and microbial behavior has produced a qualitative understanding which can explain most observed behavior. In the absence of explicit modeling methods, this understanding should enable development of a 'knowledge based' model or decision tree which gives a structured approach to investigating risks and resolving problems.
- Some processes involved in biofilm growth are not sufficiently well understood to enable accurate modeling. This includes:
  - Predation
  - Disinfection penetration and kill in biofilms
  - Sorption of substrates into biofilms producing reserves which buffer the biofilm from variations in the bulk water concentration of nutrients
- Biofilms are highly complex. The type and amount of extracellular polymers, density, thickness, viscosity, and geometric shape all influence the transport and diffusion of nutrients to the attached bacteria. Accumulation of metabolic waste products may inhibit growth unless these compounds can be released from the biofilm. Shear forces generated by the flow rate and surface topography are important factors in the release of bacteria from the biofilm surface. Characterizing the behavior of biofilms succinctly but effectively is a difficult problem for model developers.
- Several models have already been developed for microbial behavior. However, they involve parameters that are not routinely used or recorded by utilities. If utilities are to use these models then it is of fundamental importance that there is guidance on how they can determine the value of these parameters.
- The concentration of substrate, disinfectant, and pipe material are all highly important to microbial growth, but vary with location in the system. Developing methods for defining or simulating all these variations will be a challenge for researchers.

CHAPTER 4
MODELING DISTRIBUTION NETWORKS

This Chapter reviews the application of models to real distribution networks. The water quality models presented in the preceding chapters have a number of limitations in this regard:

- They generally do not account for how the hydraulic behavior propagates the water quality components through the system and how mixing and turbulence influence the reactions that occur
- Some only address reactions in the bulk water and do not account for pipe wall effects
- Where distribution system effects are accounted for, they are represented by a single set of model coefficients. Hence, they are represented as a single pipe or tank of uniform properties. In reality, distribution networks consist of many different pipe types, each with different properties and probably requiring a separate set of model coefficients.

In a few circumstances, the above may not be critical and it may be possible to give a reasonable approximation by predicting the water age and using this as an input to the 'single vessel models' described in the preceding chapters. Such an approach is most likely to be suitable when:

- The water quality problem is predominantly a bulk water issue and not significantly affected by the distribution network. This will sometimes be a reasonable assumption for THMs, but it would not be applicable to disinfectant decay or microbial water quality.
- The distribution network has similar properties throughout its entirety. This will often occur when designing new systems. It may also be relevant when analyzing a single element of the system such as storage tanks.

In most circumstances, the above conditions will not apply and so the modeling method needs to take account of hydraulic and network effects. There are several different approaches to this which are reviewed in the following chapter:

- *Zone level modeling:* Some modeling techniques are applied at a system level. Although they consider network properties, they do not have a point predictive capability beyond the zonal level. They can be used for comparing the relative performance of zones to identify areas in which control measures are required.
- *Storage models:* In some circumstances, water quality can be strongly controlled by storage. This is most notable for parameters which are controlled by bulk water reactions and where storage times are high. However, in many circumstances, it is the combined effect of the network and storage which controls water quality.
- *Network models:* These models simulate the entire network. The models incorporate simple storage models so that the combined effect of storage and the network can be considered.

ZONE LEVEL MODELS

Utilities are increasingly required to use a structured approach to justify investment decisions. This often involves methods for quantifying the current and future performance of different parts of their network. Models can be used for this assessment. However, a model with a detailed point predictive capability may not be appropriate, either because the output is too detailed for the decision making process, or because it would require excessive modeling effort. More commonly the models will be aimed at a zone level covering a discrete geographic area such as the area fed by a given source or a leakage monitoring zone. Such models can be used to:

- Identify high risk areas in which additional care is required when undertaking network operations.
- Prioritize maintenance activities
- To justify the need for renovation work

Statistical Models

As previously discussed in Chapter 3, statistical models are particularly suited to microbial parameters and can be applied directly to entire distribution networks. A major limitation of statistical models is the need for large data sets from which to derive the statistical relationships when looking at low frequency events such as microbial non-compliance. For this reason it has only been successfully applied at a supply zone level.

Tabulated Models

Tabulated conceptual models define the causes and solutions to a problem by a set of interlinked tables. The solution is derived by defining the problem parameters and working through the tables. The outputs typically will be an assessment of risk or likelihood of a condition. The major difference between this approach and most other numerical models is that the predicted behavior is not defined by a continuous mathematical function. Instead, the tables split up the model inputs, internal parameters, and output into discrete ranges, giving steps in the output. These models could be considered to be midway between conceptual and numerical.

Tabulated models are useful where the factors which influence a water quality phenomenon are well known, but where the mechanisms involved are so complex that it is difficult to produce an explicit model. Tabulated risk models allow the influencing factors to be reviewed in a systematic manner so that some indication of risk of unacceptable water quality can be ascertained. These models are often used to rank areas in order of susceptibility to a problem rather than to predict the actual risk of failure at a specific location. The approach has been used in the UK to assess discoloration risks, but similar techniques would be appropriate for assessing microbial susceptibility.

STORAGE MODELS

Storage will often have a detrimental affect on water quality due to the resultant increase in water age. In some circumstances, water quality problems will be clearly associated with a specific water storage facility in which case it can be appropriate to build a dedicated storage model,

However, in many circumstances, the interaction between the storage and the network is important, in which case it may be better to build a network model which incorporates a model of the storage facility.

A limitation of the storage models incorporated in network models is that they are highly simplified 'black box' models which simulate the storage as a whole and so cannot identify within the storage where the water quality problems occur. Also they cannot predict how the three dimensional design of the storage, such as inlet location or baffles, influence water quality. Although they might be able to simulate the overall effect of reasonably complex mixing patterns through the use of empirical multi-compartment storage models (see *Empirical Modeling of Storage Tanks* in this chapter), they will not predict what mixing patterns are likely to occur, and so the modeler either needs a knowledge of the expected behavior or to use more detailed storage models in conjunction with the network model.

Guidance on modeling storage is given in the AwwaRF report *Water Quality Modeling of Distribution System Storage Facilities* (Grayman et al. 2000). The report promotes the use of an integrated monitoring and modeling strategy. Monitoring can quantify the existing performance and produce data with which to calibrate models. Models can be used to study mixing phenomena which cannot be observed directly and used to investigate alternative designs or operating policies. Three different types of model are considered in the report: physical models, CFD models, and empirical models.

Physical Storage Tank Models

Grayman et al. (2000) found that physical models with scales of between 1:30 and 1:100 could give good results providing scaling laws are followed to maintain similitude in the predominant hydraulic forces (ratio of inertial to gravitational, viscous, or surface tension forces). This requires that flow velocity and time have different model scales to the length dimensions. Physical models can be built using simple construction methods from inexpensive materials, such as:

- 20 mm thick plywood and 50 × 100 mm timber to form the main structure
- Tile board or fiberglass board to line the tank
- Silicon sealant to seal joints
- Domestic plumbing fittings and garden hose to replicate pipe work, inlets, and outlets

The behavior of the models can be observed by adding either a chemical or a dye tracer. Suitable chemical tracers include fluoride, calcium chloride, sodium chloride, and lithium chloride. These can be measured using conductivity probes or ion specific electrodes. Patterns from dye tracers can be recorded using a camera or video mounted above the model. Grayman et al. (2000) found that dyes were good for giving a qualitative understanding of mixing behavior, whereas chemical tracers were better for numerical quantification of mixing performance.

CFD Modeling of Storage Tanks

CFD can be used to analyze flow paths through tanks and thereby identify zones with inadequate turnover of water. An overview of CFD theory is given in Appendix B. The typical steps when using one of these packages to model a storage facility would be:

1. Define the geometry of the structure and generate a computational mesh for the analysis. In many packages this can be done automatically by importing computer aided design files. Grayman et al. (2002) recommended that the optimum mesh set up was 12 to 15 cells vertically and 60×60 cells horizontally (40,000 to 60,000 cells in total), providing there were no internal pillars or barriers.
2. Define the boundary conditions for the model
3. Run the CFD solver to perform the analysis. Even with modern computers this may take several hours.
4. Display the results and review the flow paths. The location of zones in which there is poor mixing and slow turnover of water. It also helps engineers visualize the flow behavior giving them a better idea of any remedial designs to resolve problems.
5. Plot a retention time distribution curve (volume fraction versus retention time)
6. Repeat the exercise, if appropriate, to analyze alternative modes of operation or the effect of adjustments to the design; such as, the installation of baffles

Grayman et al. (2002) concluded that CFD offers 'significant potential' for simulating mixing in tanks. It allows quantification of velocities, tracer concentration (and hence water age), and disinfectant residuals at any point in the tank. CFD also allows the modeling of thermal buoyancy, although this is not generally significant unless stratification conditions occur. However, simulation times could be high (days or weeks) and the software is expensive.

HydroTank is a simplified CFD tank analysis software which was developed as part of the AwwaRF project by Grayman et al. (2000). The software is intended for use with no specific training and enables simple model set-up. It can simulate cylindrical, rectangular or hydro pillar tanks of variable dimensions. The inflow and outflow can be located anywhere below the water surface provided it is flush with and perpendicular to a wall. The mesh is automatically generated and the model output gives mixing statistics and simulated tracer curves. The main limitations of HydroTank are that:

- It is still restricted by high simulation times
- It cannot model complex geometry such as internal baffles
- The graphical output is limited compared to other proprietary CFD

Empirical Modeling of Storage Tanks

Empirical models of storage are primarily based on relationships between the observed behaviors of the tanks without regard for the underlying physical processes. There are three basic types of empirical model which are illustrated in Figure 4.1:

- *Fully mixed:* This assumes that as soon as water enters the tank it becomes fully mixed with the body of water within the tank. Throughout the tank the water will be of uniform quality and age. This is commonly called a 'continuously stirred tank reactor' (CSTR).
- *Plug flow:* This assumes that the water progresses along a fixed path from the inlet to the outlet with no mixing. In this model, the oldest water is that at the outlet.
- *Short circuiting:* This is similar to a plug flow model, but with a connected inlet and outlet. Again there is no mixing and so the water drawn from the tank is always the youngest water in the tank.

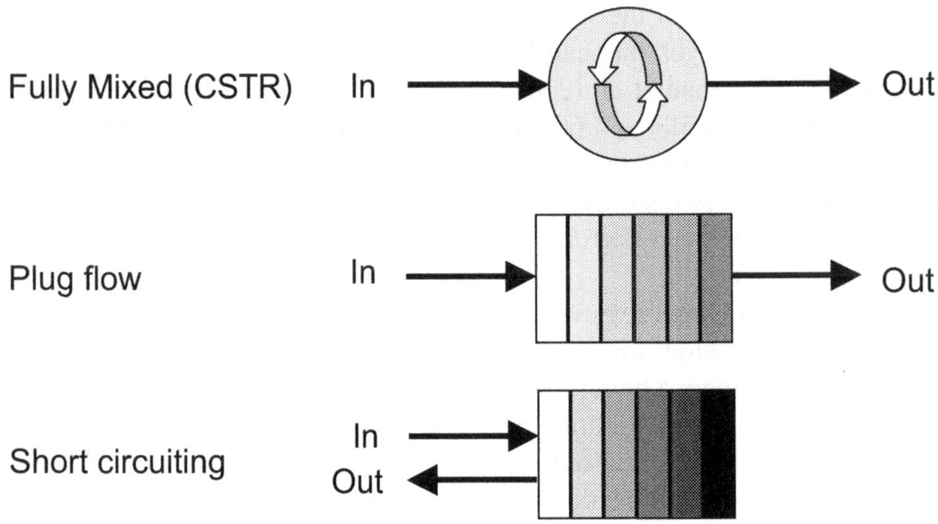

Figure 4.1 Basic empirical storage models

A CSTR will often give a simulation of the behavior of fill or draw tanks, providing that the inlet jet has reasonable momentum and there are no internal baffles or other obstructions. Whereas, contact tanks and settlement tanks at treatment plants are often designed to promote plug flow.

More complex behavior can be simulated by producing multi-compartment models in which the storage is divided into zones, each of which is simulated using one of the basic models. By doing this it is possible to simulate varying degrees of mixing, short circuiting, and dead zones. It is also possible to simulate stratification. EPANET (version 2) incorporates several different multi-compartment models for tanks. For further details on multi-compartment models the reader is referred to Mau et al. (1995) and Grayman et al. (2000).

CompTank is an empirical tank analysis software which was developed as part of the AwwaRF project by Grayman et al. (2000). It can model tanks using a variety of empirical models including multi-compartment models. The main advantage of empirical models over CFD or physical models is that they are very quick to set-up and run. This makes it viable to test many different tank set-ups and flow scenarios. The limitation of the approach is that the models use non-measurable parameters which can only be derived by calibrating them against observed data or data obtained from other modeling methods.

NETWORK MODEL THEORY

Network models attempt to define the hydraulic behavior of distribution networks by mathematical equations to be solved by computers. Many of these models incorporate some water quality modeling capabilities. If accurate, they will allow the user to investigate the behavior of some water quality parameters at any point in the network without having to go into the field to measure them directly. They also allow the user to consider what would happen if different conditions were imposed on the network; for example, to investigate how a network is likely to perform with the increased demand for water from a new housing development. Network models used to predict hydraulic behavior are widely adopted within the American and European water industries. Over the last twenty years there has been interest in the use of network models to model water quality parameters and in particular chlorine. There are numerous software packages available for network modeling, but one of particular note is EPANET. EPANET was developed by the US Environmental Protection Agency and can be downloaded for free from the internet (USEPA 2002). Since the code can be freely accessed and edited, it is frequently used by researchers as a platform for further development.

This report does not give full guidance on the extensive subject of distribution network modeling. The aim in the following sections is to explain the limitations of these models with respect to the simulation of water quality. For more general guidance, the reader is referred to the books by Walski, Chase and Savic (2001); Walski et al. (2003); Clark and Grayman (1998); and WRc (1989).

Hydraulics

Network models represent distribution networks as a set of nodes connected by links. Normally a link will be a stretch of pipes with similar properties (diameter and roughness). Depending on the software package network components such as pumps, tanks, or valves may be classified either as nodes or links, but this makes little practical difference to the use of the model. The properties for every component in the model must be defined by numerical parameters, so that the network behavior can be mathematically calculated. These parameters include:

- Pipe length
- Pipe diameter
- Pipe roughness
- Node elevation
- Demand (flow abstracted) at each node

The hydraulic analysis is based on the following two principles:

1. The sum of flows into a node equals the sum of flows out of a node. This is an application of the principle of conservation of mass.
2. In any loop, the sum of all head losses is zero

Based on the above a set of simultaneous non-linear equations can be set up for flow and head loss which are solved using iterative techniques. The Hardy Cross method (cited in Lin 2001) is widely credited as being the first systematic technique for solving pipe flow in looped networks. This method is still commonly used for manual calculations, but network analysis software often use slightly different techniques which are more efficient for computerized calculations on larger networks. There is some variation between models in the way this is done (Brendan and Kuczera 1997).

To model a system over a period of time the above procedure is used, but it is applied successively, typically at a time interval of one hour. This results in a 'sequential steady state' or 'extended period simulation' (EPS) model. A true dynamic model would simulate transient pressure waves (surge or water hammer) which are much more complex to model, but are not relevant to most water quality modeling applications. However, there has been some concern about the

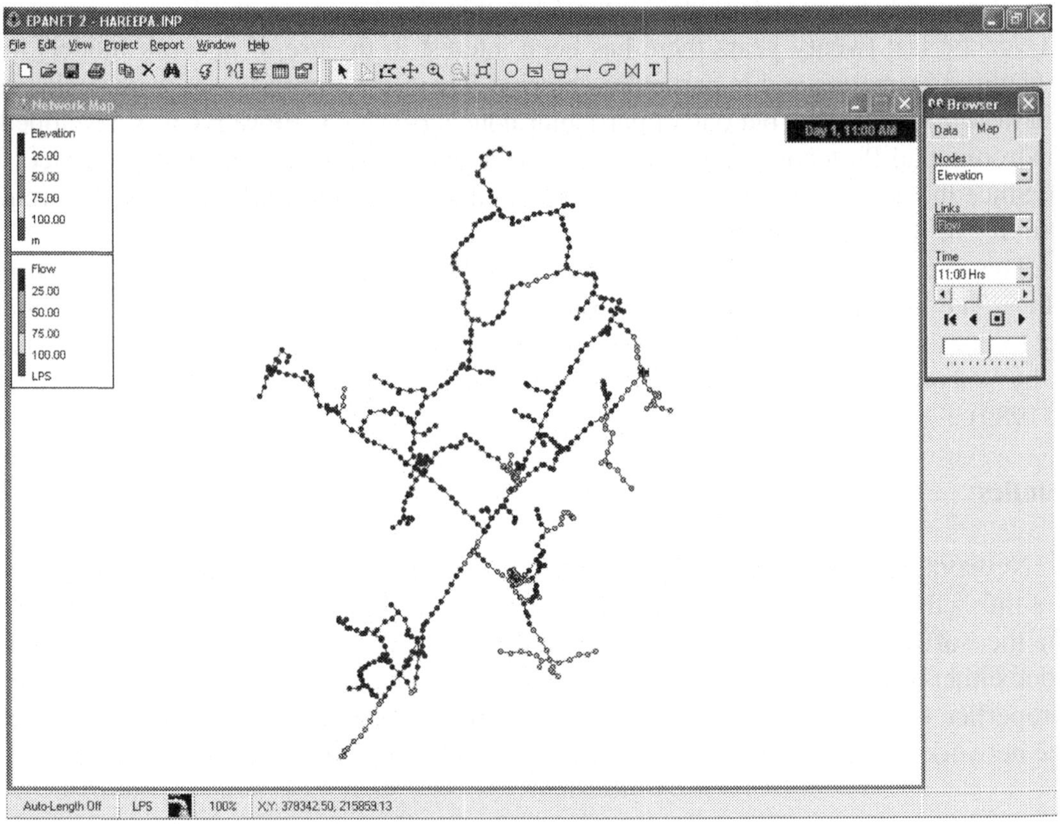

Figure 4.2 EPANET graphical display

potential for negative pressure transients to cause intrusion of microbial contaminants (McMath and Casey 2000, Kirmeyer et al. 2001).

Network models invariably produce a large amount of output data. Most modern modeling packages incorporate a graphical representation of the network which displays output data and can be used to edit the properties of individual nodes or links (Deininger et al. 1992). Figure 4.2 shows the user interface for the EPANET modeling package. Model results can generally be viewed via annotated maps, tables, graphs, color coded plots, or contour plots.

Age of Water

Many hydraulic network modeling packages incorporate algorithms to calculate the age of water at any point in the network (Liou and Kroon 1987, Boulos et al. 1994, Boulos et al. 1995, Rossman et al. 1993, Reddy et al. 1995). Models that are used to predict water age are frequently referred to incorrectly as "water quality models." Nevertheless, calculating water age is a fundamental first step when using models to predict water quality.

Unlike flow and pressure, which are only affected by the current conditions imposed on the network, the age or water quality of any parcel of water will depend on the integrated effect of the flows experienced by that parcel during preceding time-steps since entering the network. Hence, water quality models have to trace individual parcels of water as they pass through the system. This requires the model to perform the analysis at a much greater temporal and spatial resolution than is necessary for hydraulic models, thereby greatly increasing the computation required. There are various methods by which age can be calculated. Work by Rossman and Boulos (1996) compared the commonly used methods and concluded that while some were slightly more efficient than others, their numerical accuracy was very similar.

Water Quality

There are three capabilities of age of water network models which allow them to be readily adapted to predict some water quality parameters:

- Using the model algorithms used to predict water age, they can track the propagation of a slug of water (and its component water quality) as it passes through the system
- Where multiple flow paths join, the mixing of the water can be accounted for by assuming that the water quality immediately downstream of the connection is a flow weighted average of the inflowing water qualities. The models and assumptions do not generally account for any reactions that arise from the mixing of incompatible water from different sources, for example precipitation reactions which could lead to discoloration.
- The age of water predictions can be extended to predict water quality parameters, where those parameters can be defined as a function of time. The water quality models available within the EPANET software are summarized in Table 4.1.

General Network Model Limitations

Hydraulic models have been widely used to assess the hydraulic capacity of a network. The performance of models at predicting low flow conditions and the full diurnal cycle of flows is much less critical for this application than it is for water quality modeling. Low flow conditions with their corresponding long travel times are of great significance to water quality parameters. This difference in application means that there are often limitations when existing models are used to assess water quality. Methods for addressing some of these limitations are discussed in the following section of this chapter on *Network Model Development*:

- *EPS:* Where fire flows are significantly higher than the normal operational flows, a steady state model may be adequate for assessing hydraulic capacity. However, for water quality modeling an EPS model is essential. Although a steady state model can be switched to run in EPS mode, the results will be highly unreliable unless the demands and demand profiles have been allocated carefully and the model calibrated for EPS conditions.
- *Skeletonization:* Models are sometimes simplified by omitting small diameter mains. This practice, known as skeletonization, arose from limitations in computing power and the wish to reduce the effort of developing a model. In looped networks that contain multiple flow paths to some points in the networks, omitting some pipes can remove some of the possible flow paths. This will falsely increase the predicted flow and velocity along alternative routes and alter the predicted flow patterns. The use of GIS for model development has reduced the effort of constructing 'all mains' models and so skeletonized models are becoming less common.

Table 4.1
Bulk decay models available in EPANET

Equation		Comments
$\frac{dC}{dt} = k_b(C_L - C)C^{(n-1)}$	(4.1)	*EPANET bulk decay equation:* It can be set as several well-known kinetic models by adjusting the value for C_L, k_b, and n
$\frac{dC}{dt} = k_bC$	(4.2)	*First order decay* (similar to Equation 2.2): ($k_b < 0, C_L = 0, n = 1$): Widely used for chlorine decay
$\frac{dC}{dt} = k_b(C_L - C)$	(2.21)	*First order saturation growth:* ($k_b > 0, C_L > 0, n = 1$) Typically used to model THM growth
$\frac{dC}{dt} = k_bC(C - C_L)$	(4.3)	*Two-Component, Second-Order kinetics:* ($k_b < 0, C_L \neq 0, n = 2$) Can be used to model the decay or growth of a compound when the reaction rate decreases as the concentration of the compound to which it is reacting decreases. May be applicable to chlorine reactions between chlorine and NOM.
$\frac{dC}{dt} = \frac{k_bC}{C_L - C}$	(4.4)	*Michaelis - Menton kinetics:* ($k_b \neq 0, C_L > 0, n < 0$) This rate equation is often used to describe enzyme-catalyzed reactions and microbial growth. It produces first-order behavior at low concentrations and zero-order behavior at higher concentrations.

Where C = concentration
C_L = limiting concentration
k_b = bulk water decay coefficient
n = order of reaction
t = time

Source: USEPA 2002.

- *Pipe diameter errors:* The actual internal pipe diameter may be significantly smaller than the nominal diameter for the pipe due to pipe linings or corrosion. If the diameters in a model are over-sized, then the velocities will be under-predicted and age of water over-predicted.
- *Looped networks:* In networks with a dendritic (tree-like) structure, the modeled flows at any location are dictated by the modeled demands downstream of that point. However, in looped networks, the flow routing is dictated by a balancing of head losses within each loop. If the length of a loop is short, the head loss along it will be low and so there is reduced confidence in the predicted flow routing.
- *Demand allocation:* It is commonly assumed that all the domestic properties within a defined area have identical demand patterns (see section on *Network Model Development* in this chapter). In practice, there will be variability in the demand

patterns and quantity of water consumed by each property (Edwards and Martin 1995). This problem will increase greatly with the size of the area.

- *Time averaging:* EPS models are based on analyses typically at a one-hour or 15-minute time interval. The flows predicted by the model are therefore average flows across each time step. In practice the actual peaks in flow may be of shorter duration. Most real demands are either on or off (square wave demands) and pressure dependant. It is only when the demands from a significant number of properties are aggregated that they tend to result in a smooth overall demand profile that may be relatively consistent from day to day. Towards the periphery of the network, the number of downstream properties decreases and hence the actual flows become more erratic and less predictable. Hence, network models are likely to have lower accuracy in the velocities predicted towards the limits of the system.
- *Axial dispersion:* There has been some concern that axial dispersion could mean that the propagation of a chemical within the water is not accurately modeled by tracking a slug of water traveling at the mean axial velocity. This is due to:
  - The non-uniform velocity profile across a pipe which means that flow at the centre travels faster than flow at the ends
  - Chemical diffusion both axially and radially
  - Eddies and recirculation at pipe bends, fittings and other obstructions

  Axworthy and Karney (1996) demonstrated that axial dispersion was most significant at low velocities and gave guidance for assessing whether it is likely to be important.
- *Surge:* Most models are unable to replicate surge effects which can result in short periods (seconds) of low or negative pressure. Although networks should be designed and operated so as to prevent negative pressures, it has been speculated that this could be a cause of microbial intrusion and contamination of otherwise well-managed distribution systems (McMath and Casey 2000, Kirmeyer et al. 2001). Surge could also influence discoloration by disturbing sediment deposits.

NETWORK MODEL DEVELOPMENT

The model development process involves defining the value of the input variables for the model. The variables represent the properties of all the components within the network. If the properties are unknown, then assumptions need to be made for estimating their value in order for the software to calculate a solution. The accuracy of these estimates is the fundamental factor which determines the accuracy of the final model. Table 4.2 lists the principle input parameters which are required for each component in the network.

Initial Model Development

The initial model development involves a number of specific tasks. Since an EPS model is essential for most water quality modeling applications, the following sections refer to the development of EPS models unless stated otherwise.

Table 4.2
Typical input variables defined during the model development process

Component*	Parameters	Data source†	Comments
Network geometry	• Connectivity • X, Y coordinates • Node elevation	• Maps • Plans • GIS	• Connection details at junctions can be unclear • Beware that records (especially GIS) may contain errors
Pipes	• Internal diameter • Length • Roughness	• Maps • Plans • GIS • Typical values	• True internal diameters allowing for linings and corrosion may be significantly less than nominal diameters • Roughness is normally based on typical values for the pipe material, but is the pipe material and its age known?
Valves	• Loss coefficient • Degree of throttle	• Assumed values	• Difficult to assess • Often only key valves are modeled, the losses from other valves being incorporated into increased pipe losses
Storage	• Dimensions	• Drawings • Site survey • Field testing	• Need to measure and define how the storage volume varies with depth but this can be difficult
Pumps	• Pump curve	• Manufacturer • Field testing	• Pump performance will deteriorate with age due to wear
Demands	• Consumption • 24 hr profiles	• Flow metering	• Meter errors and under-recording can be a significant issue • 24 hr profiles will vary significantly across networks, the accuracy of models is particularly sensitive to the determination of representative profiles
Control rules	• Pump & automated valves control logic	• Consultation with operators	• For example if a pump switches on or off when the water in a tank is at a given level, this can normally be defined by logic rules
Water quality coefficients	• E.g. decay coefficients	• Measure • Relationships with surrogate parameters • Assumed values	• The value of some water quality coefficients can vary by several orders of magnitude, making it unreliable to use assumed values. Where possible, it is preferable to measure them directly or by using predefined relationships with other measurable surrogate parameters.

* This is not a comprehensive list. Different software may define properties in different ways.
† Local knowledge should be sought for information on all the components modeled.

Digitizing the Network Layout

This process defines the XY co-ordinates of the network, the connectivity, and the pipe lengths. The model can be digitized by using various techniques for tracing mains record drawings. This can be a laborious task, however, where mains records are held on GIS, it is often possible to export the network layout data into a format which can be imported into the model. This greatly reduces the time involved, although a significant amount of 'model cleanup' is often required particularly associated with the connectivity of pipes. Where records (either paper plans or GIS) show pipes crossing or in very close proximity, it can be unclear whether they are connected. This can be a source of serious errors in the model.

Assigning Node Elevations

The elevation at each node in the model can be assigned manually by reviewing maps, but it is quicker and often more accurate if done by electronic interpolation of digital mapping data.

Define Properties of Model Elements

It is necessary to define the properties of all pipes, storage, pumps, and valves. Pipe roughness can not be readily measured and so typical values are normally given based on the assumed pipe material and age.

Demand Allocation

This is the process by which all points of water extraction from the network are assigned a 24 hr 'demand profile' for their water consumption. The process must account for the spatial variation in the magnitude and profile of the demands across the network. Accurate demand allocation is particularly important for water quality models. With hydraulic analysis, the modeler is principally concerned with the peak flows. Whereas, for water quality analysis, the flow at every time step is of equal importance.

There are three principal demand categories which are normally assigned separately:

- *Non-revenue water:* This includes leakage, wastage, and other non-revenue uses and losses. It can be estimated by various methods (Walski, Chase and Savic 2001).
- *Industrial demand:* Large commercial and industrial users are normally measured and assigned individually
- *Domestic demand and small commercial users:* Ideally the demand profile would be recorded individually for each property connected to the network. However, even where all customers are metered, logging the 24 hr profile for each meter is unlikely to be feasible. Hence, it is invariably necessary to make some assumptions about the temporal and spatial distribution of domestic demand.

Several different methods for allocating domestic demand can be used depending on the type and quantity of data available. These methods have different benefits and limitations which are discussed below. The methods can be combined to improve the accuracy of the model.

1. *Domestic metering method:* Where domestic consumers are metered, billing records can be used to give the average daily use for each consumer. It is sometimes possible for the modeler to assign each customer to a node so that the nodal demand is based on measured data. However, it is often difficult to extract the necessary data in a format that can be readily assigned to individual nodes. It is more common to count the number of properties fed by each node. The metered customer consumption is then aggregated by area (e.g. by zip code) and allocated to each node in proportion to the number of properties. In principle basing the demand allocation on meter records should be an accurate method. In practice, there can be a number of limitations:
   - Volumes of consumption from billing records will not give the temporal demand patterns. Various studies have tried to characterize the patterns from different types of property and residential categories (Mayer et al. 1999, Clarke et al. 1997, Buchberger et al. 2003). These can be a useful basis for estimating demand profiles if no site measurements are available. However, in many areas the actual profiles show considerable localized and seasonal variations (Edwards and Martin 1995) and so characteristic reference profiles will only give a crude estimate.
   - The 24 hr demand profiles used in the model will normally be based on logged data from flow meters into the zone rather than measurements at consumer meters. However, if the zone is large (greater than 2000 properties), there may be significant variability in the actual demand profiles across the zone. This is due to greater diversity in the characteristics of the properties and the increased opportunity for extreme values. Although the demand profile for large zones is more likely to concur with the system-wide demand characteristics, this simply hides the fact that within the network, the actual demands deviate from this system-wide average.
   - Data from billing records will give the average consumption across a relatively long period and so will not give the actual spatial demand distribution for a specific day. Hence, any seasonal changes in demand level can only be assessed at the zone level. In addition, the demand data will not be specifically concurrent with data collected for calibration purposes.
   - Customer meters are prone to under-reading and their performance will deteriorate with age
2. *Zone metering method:* If the distribution network is divided into small metered zones, then the mass balance of flows and non-domestic demands can be used to infer the domestic consumption and 24 hr demand profile. These metered zones are often a permanent feature for the management of the network, although the method can be based on the use of temporary meter installations. In the UK where customer demands are not normally metered, all networks are subdivided into district meter areas (DMAs) for the specific purpose of leakage management. All flows into the DMA are recorded along with any exported flows to neighboring DMAs. Each DMA typically serves between 500 and 3000 properties. Data from all DMA import and export meters is logged and recorded, typically at a time interval of 15 minutes or less. Leakage is estimated by subtracting known and assumed values for legitimate consumption from the minimum night flow. The overall domestic demand for each DMA is calculated as the net recorded inflow from the DMA meters, minus the estimated leakage and minus the industrial consumption which is normally metered individually. The average consumption is then assigned

to each node within the DMA in proportion to the number of properties. This method allows individual temporal demand profiles to be established for relatively small areas. This allows more accurate prediction of peak flows and temporal flow patterns which are particularly important for water quality analysis. Furthermore, demand profiles can be developed which are concurrent with flow and pressure data collected for calibration purposes. The limitations of this approach are:
 - Accurate results are dependant on using relatively small metered zones (less than 2,000 properties)
 - Within each zone, the spatial allocation of demand is not directly measured, although this is little different to the domestic metering method if consumption is aggregated by area to simplify the process

3. *Estimate based on area characteristics:* If domestic customers are not metered and the network cannot be subdivided into small metered zones, then it may be possible to perform a crude demand allocation based on socio-economic characteristics for each area, such as number of residents, house lot size, household income, property value, and presence of water using appliances (Alegre and Coelho 1993, Clarke et al. 1997). This method does not measure any flows directly and so can only be regarded as an estimate and will not give a 24 hr demand profile. This approach may be of some value for refining the spatial demand allocation when using the zone metering method, especially if the zones are large (more than 2,000 properties).

It should be noted that since the fire flow requirements in the USA are particularly high, the hydraulic capacity of a network is likely to be determined by fire flows rather than other system demands. In which case, where models are only used to investigate hydraulic capacity, accurate allocation of domestic demand is less critical. However, when modeling water quality, fire flows are not normally relevant and so accurate allocation of other demands becomes much more important.

Model Calibration

At the end of the initial model development, the user should have a model which will solve numerically. However, the initial model is based on assumed values for the many parameters which are not directly measurable, such as pipe roughness or certain water quality coefficients. Until the model has been calibrated, there can be little confidence in its output. Calibration is achieved by comparing model predictions with field measurements and adjusting the value of model parameters to improve the fit between predicted and observed data. The benefits of calibration are:

- It improves the predictive capability of the model
- It tests the data and modeling assumptions used when developing the model and so indicates the confidence in the performance of the model
- The modeler can investigate the sensitivity of the model to different model parameters and in doing so will gain an enhanced understanding of the behavior of the network and any hydraulic limitations

The parameters adjusted during the calibration should be those in which there is greatest uncertainty and so an understanding of potential errors is important (Engineering Computer Applications Committee 1999). Generally the modeler will start by adjusting the parameters which can not be accurately measured and for which assumed values have been used in the model development such as pipe roughness. If unexpectedly large model adjustments are required to calibrate a model, they should be highlighted as anomalies and prioritized for further investigation. For example, valve status checks might be carried out to investigate localized areas of high head-loss.

Hydraulic Calibration

The parameters most commonly adjusted during hydraulic calibration are the pipe roughness, valve status (degree of throttle), and demand characteristics. Pipe diameter can also be adjusted, although this is less common. Head-loss is very sensitive to diameter and small errors in actual diameter can lead to significant errors in head-loss predictions. It is fairly common to use nominal diameters for pipe size and these can be significantly different from actual diameters, especially if the pipe is lined or has corroded.

The calibration process effectively checks whether the model accurately predicts the observed head-loss between measurement points. Hence, the test should be set up so that there is sufficient flow to induce enough head-loss to be measured accurately. There are two different approaches to this:

1. *Monitoring of ambient flow conditions:* If the diurnal flows are sufficient to cause significant head-loss then the field test can be based on the measurement of concurrent flow and pressure data. This is the common practice in the UK where diurnal velocities can be relatively high and the field test is used to collect not only data for calibration, but also the flow data required for the demand allocation (see *Demand Allocation – Zone metering method,* p. 49). Flows need to be recorded at all sources into the system and at internal zone flow meters. The inlets and outlets of all storage and the usage of the larger metered consumers should also be recorded. Pressures are measured simultaneously typically at between 20 and 35% of the number of demand nodes.
2. *Monitoring of artificially stressed flow conditions:* If the typical peak diurnal flows are well below the hydraulic capacity of the pipe, then there may be insufficient velocity in the pipe to cause significant head-loss under normal diurnal conditions. To overcome this problem, fire hydrant flow tests can be conducted. Pressure is recorded at appropriate locations upstream of one or more hydrants which are opened and their discharge flow rate recorded. Based on the difference in head-loss between monitoring points as the discharged flow is increased, it is possible to calculate the roughness between monitoring points. The benefit of this method is that it can test greater extremes in hydraulic conditions. The limitations of this method are that:
   - It does not record the diurnal pressure fluctuations which can give an insight into the behavior of the network
   - It will not indicate anomalies with the ambient flow patterns and so will not validate the assumed demand allocation

- Many tests are required to cover the whole network comprehensively. For this reason the modeler may try to isolate a set of characteristic pipes in this way and assume that each category of pipe has consistent properties throughout the network.

The quality of a calibrated model is directly related to the quality of the data derived from the field test and the number of reference points monitored. Increasing the number of reference points will also help identify anomalies and can reduce the calibration effort. Confidence in the calibration can be improved in a number of ways:

- There is merit in performing a combination of the two above monitoring methods, especially when developing models for water quality simulation where it is important to test the model's performance across the full diurnal flow cycle, not just the performance under peak flows
- All calibration adjustments should be clearly documented and any significant anomalies investigated and rectified or simulated in the model. Anomalies are often caused by major leaks, closed valves, or unknown connections between pipes and unrecorded significant demands.
- Prior to the calibration, the model should be set up to reflect the supply and demand conditions which are contemporary with the calibration data. This will require different demand profiles and flows to those used in a final model which would normally represent average day or peak day flow conditions.
- In addition to matching pipe flows and pressures, the model should replicate storage tank, pump station, and valve performance. The calibration should test the overall system performance. It is preferable to establish pump and valve performance curves from in-situ testing rather than to rely on manufacturers' data.
- Tracer studies (see below) can be used to determine flow paths and flow rates
- The diurnal flow patterns can show considerable variability from day to day. It is normal for weekends to have different demand patterns compared to midweek, but smaller variations may occur within the week. Monitoring flow and pressure over several days or, ideally, for a full week allows the modeler to assess the consistency of the network behavior and ensure that representative data is selected for use in the calibration.
- Altering nodal demands or diurnal demand profiles as part of the calibration process is normally inappropriate unless it can be justified by logical reasoning. This is for a number of reasons:
  - It is difficult to determine error bands within which it is reasonable to adjust demands
  - It can be difficult to accommodate calibration adjustments to demand when models are updated to represent different demand scenarios
  - It will often be possible to get a similar 'improvement' in model fit by adjusting different demands, at different nodes, in different ways. Hence, there are likely to be multiple solutions. It is the authors' experience that where models have been calibrated in this way, and the network subsequently subdivided into smaller metered zones, field data often shows that the calibration adjustments chosen were incorrect.
- In order to limit the number of unknowns before the calibration field test and thereby avoid an abortive field test, extensive preliminary field investigations are advisable.

This is particularly important when flow data is recorded for demand allocation purposes when it is necessary to ensure that all flow meters into and out of the zone are operating accurately and all metering zone valves are fully shut. This preliminary work would include flow mass balance checks, zone valve status checks, and installing additional flow and pressure monitoring points.

Water Quality Calibration

There is much less guidance available on the calibration of water quality parameters than hydraulic parameters. In the absence of such guidance there is danger that network modelers, many of whom have little knowledge of chemistry or microbiology, will treat water quality coefficients in a similar way to the calibration of hydraulic parameters. Parameters should only be calibrated if there can be confidence that their calibrated value will be more accurate than an assumed value. This will only occur when the predicted data is reasonably sensitive to the adjustment of that coefficient and if there is only one model coefficient that will have a similar effect on the predicted data. This might be reasonable for wall decay coefficients in a chlorine model, but would not be so for microbial models.

With water quality models it is often more effective to evaluate the water quality coefficients via laboratory tests or field measurements. For example, bulk decay coefficients or THM growth rate coefficients can be measured using jar tests. If this is not possible, then it may be possible to define relationships with indirect measures of the precursors to the reactions such as UV or TOC (Vasconcelos et al. 1996).

Tracer Studies

Accurate information about the flow patterns within distribution networks is essential when modeling water quality. Flow meters only give information about a single location and the number of points at which it is viable to install permanent or temporary flow meters within a network is limited by cost. Pressure monitoring can only give accurate information about flow patterns where there is a steep hydraulic gradient (hydraulic grade line). Tracer studies offer the flexibility to determine the travel time to virtually any point within the network. Tracer studies can also be used to assess the blending of different source waters within a network.

Tracer studies involve injecting a chemical into the water supply for a fixed period. The water is then monitored at downstream points to determine the duration before water containing the chemical passes the monitoring points. The use of tracers as a means of increasing the accuracy of a hydraulic model, especially when the hydraulic model is to be used for water quality modeling, is documented in the literature by Clark and Grayman (1998); Grayman (2001); Cesario et al. (1996); Powell et al. (1999); Walski et al. (2003); and Skipworth, Machell and Saul (2002). Although the principle behind tracer studies is straightforward, there are a number of details which need consideration when planning a tracer study:

- *Selection of appropriate trace chemical:* The trace chemical must be non-toxic and approved for use in drinking water. Utilities will often prefer to use chemicals which are routinely added to drinking water supplies. Ideally, the trace chemical should be easy to measure accurately at low concentrations using a portable monitor which logs the recorded data.

- *Duration of dose:* A relatively short duration is preferable because it minimizes the impact of any water quality concerns and produces a sharper trace, especially if there are multiple flow paths. However, it needs to be long enough that:
  - It is not confused by any oscillations in the ambient concentration of the trace chemical in the water
  - The spike will be picked up by the monitoring i.e. the duration of the spike needs to be significantly longer than the sampling frequency
- *Sampling frequency:* The greater the frequency of the sampling, the better the resolution of the recorded trace. In a looped network this is particularly important as a single peak at the injection point will be split into several peaks wherever there are multiple flow paths to a sampling point.
- *Sampling method:* Onsite measurement techniques are preferable because it is apparent when the trace has passed the sampling location and so sampling will not be ceased prematurely or continued needlessly. Monitors with data loggers are the most effective sampling instruments as they can be left running for long periods at a high sampling frequency. However, they can be expensive.
- *Travel time and network complexity:* For high travel times there may be problems with decay or dispersion of the trace and multiple flow paths can make interpretation of the results difficult (Benrahnem 1996, cited in Desjardins et al. 2000). Hence, it may be necessary to dose a longer or higher magnitude trace.

Despite the value of tracers in validating hydraulic models and water age calculations, routine application in distribution networks is relatively rare. A number of reasons have been suggested:

- Lack of affordable automated monitoring devices for some tracers
- High cost and effort associated with manual sampling methods
- The need for approval from industry regulators to introduce some tracers into water supplies
- Lack of studies on larger distribution systems
- Cost benefits of tracer studies in systems where flows may frequently be changing

Tracer studies in distribution pipe work generally assume plug flow of the water. However, with storage tanks mixing characteristics become important. Guidance on the use of tracer studies to assess the hydraulic characteristics of water treatment facilities is given in a previous AwwaRF report (Teefy 1996) and similar techniques can be used for distribution storage facilities. Further detailed guidance on the use of tracer studies in distribution storage is given in another AwwaRF report (Grayman et al. 2000). This second report contains guidance on the selection of trace chemicals which would be equally valid for network tracer studies. When performing tracer studies on tanks, the profile of the concentration of the tracer chemical against time indicates the mixing characteristics.

Auto-Calibration

A number of network modeling packages have been developed which include auto-calibration software. These are optimization tools based on various techniques such as genetic algorithms (see Appendix B). The optimization tools adjust the values of given model parameters until they give a mathematical optimum fit between the predicted and observed data. There are mixed opinions of the benefits and costs of auto-calibration generally associated with the subjectivity of the process and the need for practical hydraulic engineering experience. Limitations of auto-calibration include:

- It is often possible to obtain a calibration fit within acceptable tolerances by adjusting different parameters. A modeler can use his knowledge of the network and engineering judgment to make an informed decision as to the most appropriate solution. Auto-calibration can find a best fit solution, but cannot compare the appropriateness of that solution with alternative solutions.
- Model calibration often involves the identification of anomalies in the network. An experienced modeler may be able to recognize from the field data that there is a closed valve in the network or that there is an unknown connection between two pipes. Auto-calibration will not do this.
- Unless specified otherwise, the software will treat the properties of each pipe separately. A more realistic solution is obtained by grouping pipes into runs in which the properties are anticipated to be similar.
- The calibration adjustments tend to be focused in close vicinity to the location of monitoring points. The software lacks the intelligence to infer implication that similar areas elsewhere in the network may require similar adjustment and sometimes will give large localized adjustments when it might be more realistic to use smaller adjustments over a larger area.

Despite these limitations auto-calibration can be useful. Manual calibration is an iterative and sometimes lengthy process of trial and error. If auto-calibration is used carefully by controlling the variables to be optimized and reviewing alternative solutions it can reduce the calibration effort.

Model Verification

Models can only be genuinely verified if separate data is used for calibration and verification, and if the verification conditions are significantly different from the calibration conditions (Thommann 1982, Reckhow et al. 1990, Parker et al. 1995). Different hydraulic conditions can be achieved by opening fire hydrants or altering valves to substantially vary the flow pattern within a network (Cubillo and Mercier 1993). If the verification data does not fit with the model data, the data may be used for further calibration in which case a new set of validation data should be collected for hydraulic conditions which are different again. Validating models greatly increases the confidence of their predictions. However, due to the effort involved, it is commonly omitted.

USING NETWORK MODELS TO ASSESS WATER QUALITY

Age of Water Modeling

Age of water is sometimes used as a surrogate indicator of water quality. This avoids the complexity of modeling water quality. It can be a useful approach when there is low confidence in the ability to predict accurately the parameter of concern, such as with microbial water quality.

The approach relies on the basic principle that younger water is likely to be of better quality than older water. A major limitation of using water age as an indicator of water quality is that the reaction rates are often controlled by contact with the internal surface of the distribution network. The geometry, material, and condition of those surfaces will vary throughout the network and so water with a given age in one part of a network may have significantly different quality to water with the same age in another part of the network. For example, the quality of water is likely to be significantly better where it has only passed through new plastic pipes than where it has passed through a system which contains sections of old unlined cast iron pipes.

Nevertheless, age modeling can be useful as part of a screening methodology. Brandt et al. (1999) and Brandt (2000) aimed for a maximum water age of four days for some networks in the UK where the maximum chlorine dose was 0.5 mg/L. Such values for maximum water age need to be developed on a system specific basis. Lower NOM, newer pipes, or higher chlorine residuals would allow higher water ages to be safely accommodated.

Chlorine Modeling

Free chlorine has been widely modeled using network models. As previously discussed, there has been much debate about the benefits of various kinetic models for chlorine decay models, but most network modeling studies to date have assumed first order decay. This has the merit of simplicity and due to its wide-scale use most existing data on chlorine decay rates has been presented in terms of first order decay coefficients.

Modeling Bulk and Wall Reactions

A key choice for the user is whether to model bulk and wall decay separately or use a single decay coefficient to describe both. Although separating bulk and wall coefficients may appear to increase the complication, there are benefits in doing so:

- It allows the user to simulate what would happen if changes were made to the source water quality (change in bulk decay). For example, one utility successfully simulated the commissioning of granular activated carbon (GAC) treatment by reducing the bulk decay coefficient so that it was similar to that for other GAC treated water. This allowed them to avoid a sudden increase in the concentration of chlorine received by customers and thereby prevented taste and odor complaints
- The impact on water quality of pipe renovation can be simulated by reducing the wall decay coefficient. This can be useful for evaluating different solutions to a water quality problem.
- By considering wall decay separately, it is more likely that the decay coefficients for the characteristic pipe types will be consistent with values observed in previous studies of similar areas. If an overall coefficient is used, the decay will be influenced by many additional variables relating to the treated water quality and so be much harder to characterize.
- It allows the model to be updated more easily in response to changes in the network's environment such as temperature or the organic content of the treated water. These can affect the bulk and wall decay in different ways (see *Durability of Model Calibration* later in this chapter).

Modeling bulk and wall decay separately requires a limited amount of additional sampling to estimate the bulk decay coefficient for each water source. This can be measured by a relatively simple jar test. The wall decay is harder to measure directly and it is normally inappropriate to use literature values as the values reported are highly site specific. However, the wall decay may be assumed to be the overall decay minus the bulk decay. Hence, modeling bulk and wall decay separately need not necessarily be significantly more expensive than only modeling overall decay.

Parameter Estimation

One of the main sources of additional modeling effort for developing chlorine network models is the need to define decay coefficients. It is preferable to measure parameters directly rather than estimate them thereby reducing the number of unknowns.

Bulk decay: This can be readily measured using jar tests in which the water is placed in a set of sealed glass bottles which are opened at fixed time intervals and the residual disinfectant concentration measured. A plot can then be produced of concentration against time to which the selected decay model can be fitted. The test is relatively simple but there are variations in how it has been done and the method for fitting the decay model to the data. These variations could have some effect on the results obtained. The adoption of an industry standard test would help ensure consistent results.

Wall decay: Various different methods have been used for evaluating pipe wall decay, however they all have their limitations:

- *Tests on isolated pipe stretches* (Sharp, Pfeffer and Morgan 1991; Vasconcelos et al. 1996): In most networks it can be difficult to isolate individual stretches of pipe which are long enough to cause measurable decay, made of a consistent material which is characteristic of the rest of the network, and in which the flow can be directly measured.
- *Laboratory testing of cut out pipe samples* (Gotoh 1989): This can be similar to a jar test for bulk decay but with a pipe sample used in place of the jar. This ignores the hydraulic effects which are likely to be significant in controlling the radial mass transport of the disinfectant to the pipe wall.
- *Tracer studies:* Tracer studies can give the time of travel accurately and, by recording upstream and downstream chlorine concentrations corresponding with the trace, it is possible to get a reasonably accurate measure of the decay rate. However, if there is a combination of different pipe characteristics along the route of the tracer test, it can be difficult to apportion the decay.
- *Downstream series sampling* (Chambers, Creasy and Joy 1995): With this method a model is used to predict the travel of a slug of water through the network. Field sampling is then scheduled so that theoretically all samples for the disinfectant residual are taken from the same slug of water. This method relies on an accurate model and can require samples to be taken at inconvenient times.
- *General field sampling:* Most commonly a set of field samples is taken and the decay coefficients calibrated to produce a best fit against them. This is a rather indirect method of measurement which can lead to errors arising from inaccuracies in the model and assumed boundary conditions.

With all the above techniques, the bulk decay has to be accounted for by subtracting the decay observed from jar tests from the overall observed decay. There is a clear need to investigate alternatives and produce industry guidance on the most effective methods.

Durability of Model Calibration

While numerous studies have shown models to be capable of being calibrated to reproduce the observed network conditions for a given day, their performance can rapidly deteriorate. This will partly be due to changes in demand patterns or the hydraulic setup of the network which must be updated to represent the conditions being modeled. In addition the decay coefficients, in particular the bulk decay coefficients, can show significant seasonal variation. This is because the rate of decay is influenced by a number of parameters which may vary significantly with time, but which are not included in the first order kinetic model. In particular, Powell et al. (2000a) has shown the bulk chlorine decay to be controlled by temperature, NOM, and the initial chlorine concentration. These can be accounted for in a number of ways:

- The ideal method is to re-measure the decay coefficients. This is relatively straightforward for bulk decay, but would require considerable new fieldwork for wall decay. However, the wall decay is probably more stable than the bulk decay and so less likely to need adjustment.
- Updating the decay coefficients in accordance with pre-defined relationships:
  - Bulk decay can be updated assuming relationships with temperature, organic content, and the initial chlorine concentration (Equations 2.18 and 2.19)
  - Wall decay coefficients can be updated assuming an Arrhenius relationship with temperature (Equation 2.19).

  These relationships will be site specific and so some sampling and analysis will be required to evaluate these relationships, but once this exercise has been done, relatively little additional work is required for future updating of the model.
- Monitoring chlorine levels at a strategic point in the network and factoring all the decay coefficients up or down until the model gives a reasonable match. This approach is robust providing the entire network is affected by a similar change in conditions. This will normally be the case for single source networks, but may not be accurate for multiple sources in which the qualities of the source waters are not consistent.

Laboratory work has shown that some of these limitations might be avoided by using second order models (Clark and Sivaganesan 2002). There is a need to verify this by using them within a network model. This work would need to consider pragmatic methods for evaluating the additional parameters involved.

Chloramine

To date most network modeling of chloramine has treated it in a similar manner to free chlorine, but using lower decay coefficients. This has sometimes produced a reasonable fit between modeled and observed data for one moment in time. However, as discussed in Chapter 2, chloramine is different to free chlorine and first order decay is theoretically less suitable. Alternative models based on the underlying reactions involved in chloramine decomposition have been developed. There is a need to integrate such models with network models and trial their use.

THM Modeling

THMs and HAAs are regulated in the USA by the DBPR. The maximum contaminant limits (MCL) are 80 μg/L for THMs and 60 μg/L for HAAs. Fixed sample sites are used and the 'Locational Running Annual Average' (LRAA) for each site must comply with the MCLs. An Initial Distribution System Evaluation (IDSE) is under consideration. If implemented, the IDSE will require utilities to select representative sample sites, unless the water utility can demonstrate from two years of data that distribution samples are less than 40 μg/L for THMs and 30 μg/L for HAAs. There are two options for conducting an IDSE:

1. Conduct a Standard Monitoring Program (SMP). This involves an increased level of THM and HAA monitoring for one year. The data is then reviewed to determine representative sites for compliance monitoring.
2. Alternatively the utility can conduct a System Specific Study (SSS). This allows a reduced level of sampling by utilizing specific knowledge of the system to substantiate the selection of representative sites. In particular the modeling of retention time is often a key element of an SSS.

It is a notable development in the regulations that the DBPR specifically identifies the use of modeling. Some guidance on how to use models to select monitoring sites is given in a draft document from the USEPA (2003). This document includes a minimum specification for the detail of the models used, however it does not give guidance on how to develop and calibrate the models. Utilities are likely to need some guidance on the most effective modeling methods for meeting these requirements.

Some of the first applications of water quality models were directed at modeling THMs (Males et al. 1988, Grayman et al. 1988, Clark et al. 1988). These assumed THMs to be conservative, having reached their maximum potential concentration at the treatment plant. Hence, they simply modeled the mixing of water in multiple source networks where the different sources contained different THM concentrations. As such, they could be a useful tool enabling the optimization of valve settings and pump scheduling to maximize the dilution of a source with a high ambient THM concentration. The models were, however, of limited point prediction capability. This was largely attributed to the THMs not being truly conservative and rapid fluctuation of the source THM concentration.

Clark et al. (1994) demonstrated that THM growth tended to decrease with time asymptotic to a maximum formation potential. This can be modeled using EPANET with a first order saturation growth model (Equation 2.22). This has given better results than treating THMs as conservative (Vasconcelos et al. 1996).

The growth of THMs is generally modeled as a bulk water phenomenon that neglects the interactions at the pipe wall which, as previously discussed, may either increase or decrease THM concentrations. This approach makes THMs much simpler to model than chlorine as the reaction coefficients apply globally to the model and can be measured directly via a jar test with no need to calibrate separate coefficients for each pipe. This also means that accuracy will not be impaired where the THM model is not integrated into the network model. Seidal et al. 2002 described such an approach based on predicting retention times at key points and then predicting THM formation using the USEPA water treatment plant (WTP) model. They applied this to both THM and HAA formation. The purpose of the work was to identify representative compliance monitoring sites as part of the IDSE as required by the Stage 2 DBPR.

The danger of applying these empirical THM formation models is that the modeler may treat the model development as a curve-fitting exercise without an appreciation of the underlying causes of THM formation. However, mechanistic models of THM formation such as that developed by Westerhoff et al. (2002) are probably too complex for incorporation within a network model.

Discoloration Modeling

The authors are unaware of any model that fully models discoloration in water distribution systems taking into account:

- The rate of production of sediment by corrosion and other means
- The progressive deposition of sediment over long periods (years)
- The erosion and entrainment of sediment deposits including the influence of the chemical properties of the water on the stability of deposits
- The transportation of sediment

Some models have been developed which aim to indicate those regions which may be prone to the accumulation of sediment and those regions in which sediment is likely to be re-suspended. They do this by comparing predicted velocities with predicted values for the deposition limit and the threshold of motion. There are some limitations with how most existing models do this (Ackers, Brandt and Powell 2001):

- The sediment transport models which they are based on have not been validated on the conditions typically found within distribution networks (see Chapter 2)
- The models consider a single 'characteristic' particle size. However, it is possible to get sediment varying in size from less than 1 μm to over 30 mm.
- The models do not account for any cohesion or entrapment of particles at the pipe wall
- The models do not simulate the production of sediment from within the network due to deteriorating pipes
- The output from the models is a simple identification of those pipes which are supposedly at risk of suspension of sediment or deposition of sediment. This crude analysis would only have limited benefit for managing discoloration.

Nevertheless, network models can be used effectively to look for hydraulic characteristics which indicate an increased risk of discoloration such as high retention times or flow reversal. This can be useful to check whether a proposed operational activity such as shutting down a main for maintenance risks causing discoloration.

Microbial Modeling

The SANCHO model calculates the changes in BDOC, chlorine concentration, and free and fixed bacteria under steady state conditions in relation to retention time (Chapter 3). Mary-Dile et al. (2000) developed an interface between SANCHO and a hydraulic network model. Water residence time information was extracted from the hydraulic network model and used in

SANCHO to estimate the bacteriological quality of water at any given location in the network. The simulations showed that hydraulics can account for a large proportion of the variation in bacterial numbers that are reported in the distribution network. The authors also indicated that models for THM formation could be developed based on similar principles.

The developers of the PICOLLO network modeling package developed an add-on module for their software, called PICCOBIO, which models the suspended biomass through a distribution system (Piriou et al. 1997, Gagnon et al. 1997). Few details of the underlying equations are available, but the following mechanisms are apparently incorporated:

- The influence of biodegradable organic carbon on suspended and fixed biomass
- The influence of temperature on bacterial dynamics
- Bacterial mortality due to age and grazing
- Bacterial mortality due to chlorine, taking into account the effect of pH on disinfection
- Deposition and detachment of bacteria to and from the fixed biomass

The developers of the model claim a good correlation between modeled and observed viable bacteria counts ($R^2 = 0.80$) when applied to a French distribution network. The authors are not aware of any other published results from the use of this model, but understand that application in the UK has been less successful.

Contamination Modeling

The concentration of some water quality parameters will not grow or decay within distribution. Nevertheless, it can still be useful to model them using network models as it allows the user to investigate:

- The propagation of fronts of high doses
- The blending or dilution of concentrations within the system

These applications do not rely on a water quality model, although they use the same age of water calculation algorithms. Modeling conservative substances has the following applications:

- Identifying the source of contamination in multiple source networks (Clark et al. 1991, Geldreich 1991). The results of such studies have been used in legal proceedings (Clark 1991).
- THM modeling – see previously
- Blending of nitrates
- Investigating discoloration problems caused by the mixing of water from two different sources
- Operational management of pollution incidents, but only if the models include all pipes and have been maintained in a calibrated/validated state.

Real and Near Time Network Modeling of Water Quality

There is an increasing availability and use of sensors which can record various water quality parameters within distribution. Although it is clearly better to measure a parameter directly than to try and predict its value, it is only feasible to monitor at a limited number of locations and there is likely to be considerable variability in water quality elsewhere in the network. Modeling techniques could potentially be used to predict the spatial variation in real or near time.

Network water quality models have often been found to give reasonable predictions for a given day, but then rapidly fall out of calibration as they do not replicate the effect of changes in system conditions (see *Durability of Model Calibration* previously in this chapter). This is largely due to temporal variation in the factors which affect the model coefficients. If models are used as a method of spatially interpolating data between given sample points, this problem is avoided. This could be achieved by developing methods for re-calibrating model coefficients based on the current available data. Network modeling software is available which can import data from telemetry systems and optimize model coefficients so as to best fit the observed network behavior. For example, model demands can be factored up or down in proportion to changes in recorded flows and decay coefficients can be optimized to improve the fit between the current recorded chlorine residuals and model predictions. Neural networks or genetic algorithms (see Appendix B) could be effective methods for achieving this.

Combining this updating procedure with regulatory sampling could significantly enhance the interpretation of regulatory data. Regulatory monitoring is inevitably limited and there may only be one or two chlorine samples taken from each network per week. However, in many networks it is possible to find regions of high or low chlorine concentration, making it difficult to interpret results. Network models could be used effectively to interpolate that data to give a full temporal and spatial picture. Models could also be used to quantify the performance, for example by calculating the percentage of the demand containing less than 0.05 mg/L and thereby allowing direct comparison between the performances of different networks. By combining this with demand and temperature forecasting, it is potentially possible to give warning alarms before a failure occurs. Given the increasing trend to integrate network modeling with GIS, real or near time modeling is becoming feasible for many water utilities.

SUMMARY AND RESEARCH NEEDS

The following section summarizes issues related to the application of water quality models to distribution networks.

General Issues

- *Water quality skills for modelers:* Modelers need an increased and clear understanding/guidance of the multiple relationships between physical, chemical, and biological parameters and the impact of changes in one parameter on the others. This will allow them to recognize and understand the possible causes when models do not fit observed behavior.
- *Temperature effects:*
  - The scale of the impact of temperature on water quality parameters
  - Significance of diurnal temperature changes
  - Predicting and accounting for temperature changes
  - Significance of buoyancy and stratification effects in tanks and how they should be modeled

- *Integration of modeling methods:* It should not be overlooked that a problem can be approached by combining various different modeling techniques. For example velocities predicted by network models might be used as an input to a risk matrix discoloration model. Another area worthy of further research is the integration of CFD with network models. CFD could be used to model storage tanks in the network. It could also be used to model water quality interactions in characteristic sections of pipe work from which parameters could be derived for simplified models which are solved by the network model for the entire system.
- *Model robustness:*
  - Application of models to conditions outside those from which they were derived or calibrated
- *Estimating model accuracy:*
  - Assessment of the accuracy or confidence limits of model predictions
  - The feasibility of network modeling software to calculate error bands automatically for all predicted data and the additional data that this would require
- *Network design and operational management:*
  - Methods for overcoming the impact of design criteria such as fire flow capacity and operational practices on the ability to model the network and where different modeling approaches are required
- *Forecasting:*
  - Adjustment of model parameters when models are used for forecasting and how changes in these parameters can be forecast

Zone Level Models

- *Tabulated models:* There is merit in using tabulated models where there is sufficient knowledge to enable a structured approach, but where detailed numerical modeling is impractical
- *Statistical models:* Methods for the statistical modeling of very low frequency events

Storage Models

Storage can be modeled either in isolation or as part of the distribution network. If modeling in isolation, it is important to consider how the storage interacts with the distribution system and whether some of the connected pipe work is effectively part of the storage. Research needs include:

- Further development of CFD models tailored to water system tanks and reservoirs making them easier to use and faster to solve
- Develop a database of tank compartment model coefficients for characteristic tank types

Network Model Development

Network models have been used for many years and distribution system modelers have a good understanding of the issues which control the quality of the model output. The primary need is to evaluate the importance of these issues and provide guidance on specific methods and standards which should be achieved. Particular consideration needs to be given to how these methods and standards differ for different modeling applications:

- *Model development:*
  - Clarification of the limitations of skeletonization and the appropriateness of its use
  - Export of uncalibrated models from GIS and the need for model 'clean-up' tools
  - Effective use of auto-calibration
- *Quality and quantity of base data:*
  - Identification of the appropriate data quality for different modeling applications and the implications for the quantity and quality of calibration data required
  - Methods that a modeler can use to improve the quality of the demand allocation and the derivation of diurnal profiles, especially where there is limited in-line flow metering
  - Methods for maximizing the quality of data at minimum cost
- *Calibration and validation:*
  - The standard of hydraulic calibration needed when a hydraulic model is used for water quality modeling as opposed to more standard uses of hydraulic models
  - Refinement of tracer methods for calibrating hydraulic models including the development of inexpensive continuous monitors for use in tracer studies
  - The circumstances when a network model should be calibrated and then validated with a separate data set
  - Placing networks under hydraulic stress when collecting calibration data
- *Tuberculation:*
  - In-situ methods for determining the actual pipe diameter of older pipes
  - Methods for characterizing the degree of tuberculation and how it affects water quality
- *Water quality modeling:*
  - Appropriate use of surrogate parameters in water quality models
  - Whether it is appropriate to treat water quality model input variables as adjustable 'calibration' parameters
  - Simplification of water quality models to include fewer parameters
  - Simple characterization of water sources and the parameters required
  - Whether axial dispersion should be explicitly modeled
  - The need for network modeling software with the flexibility to incorporate user defined functions and parameters so as to support the development of water quality models
  - Modeling multiple source interactions

Using Network Models to Assess Water Quality

Chlorine Models

Chlorine has been widely modeled to date, and the concept is well accepted. The principle research needs centre on effective application within utilities rather than the development of new models.

- *Standard bulk decay test:*
  - A standardized method for jar tests would help ensure accurate measurement and comparability of results
- *Second order models:*
  - Testing of second order models in network modeling studies
- *Database of decay coefficients:*
  - The development of a comprehensive database of bulk and wall decay coefficients, if feasible. This will require careful consideration of the parameters required to characterize the source water and pipe type effectively.
- *Wall demand:*
  - In-situ measurement techniques for wall demand
  - Investigation into, and explanation of why, flow and velocity have been observed to alter the wall demand
  - Prediction of the effect of pipe renovation on decay coefficients

Chloramine Models

To date most network modeling of chloramine has treated it in a similar manner to free chlorine, but theoretically this is not appropriate. There is a need to apply the models of chloramine decay that have been developed from theoretical and laboratory sources to the distribution networks. In addition it would be beneficial to develop robust methods for converting free chlorine models to predict chloramine decay if the disinfectant is switched.

DBP Models

The implementation of the Stage 2 DBPR promotes the use of modeling THMs within distribution, but there is a need for detailed guidance on how modeling can be effectively used within the IDSE to expand on the information currently available from the USEPA (2003). In particular:

- There are a number of different models which can be used for THM prediction of varying complexity. Some of these were developed for treatment plant applications. Guidance is needed on the suitability of these models for predicting THMs within distribution and how they can be incorporated into network models.
- Methods for estimation of model input variables and establishing their level of consistency throughout the system.
- The consequences of blending multiple source waters within a network
- Whether it is appropriate to use similar methods for modeling THMs and HAAs

Sediment and Discoloration Models

Some network models have been developed with some functionality for predicting build up of sediment and the risk of discoloration. However, this performance is inhibited by the lack of appropriate models for the underlying processes (see Chapter 2). Nevertheless, network models can be used to check for hydraulic characteristics which indicate an increased risk of discoloration such as high retention times or flow reversal.

Microbial Models

Ensuring excellent microbial water quality is often the primary water quality concern for utilities. Unfortunately the ability to model it accurately is still limited. Nevertheless, research over the last twenty years has enhanced our understanding and the knowledge gained could be used to give a structured approach to minimizing risks. Particular research needs include:

- The simplification of models and the characterization of input variables so that they are viable for integration with network models
- Investigation of whether surge is responsible for causing microbial contamination, how significant is it as a cause of compliance failure, and the overall risk to public health

Real Time and Near Time Modeling

Real time modeling of water quality is becoming increasingly viable based on the integration of models with telemetry systems. Important issues include:

- Methods for obtaining accurate real time data and accommodating monitor drift errors
- Feasibility of basing near time modeling on manual sampling
- Methods for updating model parameters in response to the current observed data and accounting for data anomalies

CHAPTER 5
CONCLUSIONS AND RESEARCH NEEDS

This report reviews the current state of predictive water quality modeling and identifies future research needs. The review has been broken down into three topics:

1. Modeling chemical and aesthetic water quality (Chapter 2)
2. Modeling microbial water quality (Chapter 3)
3. Applying models to distribution networks (Chapter 4)

Chapters 2 to 4 give a complete review of each topic and include conclusions and specific research needs.

The first two topics have an academic and scientific focus. The third topic is more applied and has an engineering focus. In order for utilities to gain the benefits of scientific research, it is essential that the knowledge from the research scientists is transferred to modeling engineers working for utilities. Likewise, in order to steer effective research, it is important for scientists to have an understanding of the needs of utilities and the constraints which they are under. A workshop was held which brought together experienced people from both groups. Together, they reviewed the research needs and developed eleven research briefs to address the most urgent needs; these are included in Appendix A.

WORKSHOP DETAILS

The Workshop was attended by 17 delegates from the USA and UK with a broad range of expertise. The delegates included utility practitioners, academics, consultants, and representatives from the USEPA. A full list of the delegates is given in the Acknowledgements.

The workshop commenced with a series of presentations which outlined the current state of knowledge in predictive water quality modeling. Details from these papers have been incorporated into Chapters 2 to 4 of this report. Following this the delegates divided into three workgroups to review research needs in the three topic areas identified above. Each workgroup proposed a list of research topics relating to their subject area and then selected the four which they deemed to be of greatest importance. The workgroups drafted a research brief for each of the selected topics. Then the research briefs were reviewed by all the delegates at the workshop and votes taken to prioritize them. The voting was structured to ensure that at least one research brief was selected from each of the subject areas.

The research briefs have been sent to AwwaRF and shall be considered by the Research Advisory Council (RAC) for inclusion in AwwaRF's future research program. Following this review, some of the briefs may be adapted and issued as a 'request for research proposal' (RFP).

PROPOSED RESEARCH BRIEFS

The workshop attendees selected six of the briefs to be of a higher priority. These are listed below:

- *Guidelines for developing, calibrating, and using hydraulic models for water quality modeling:* Hydraulic analysis is often predominantly concerned with the peak flows and minimum pressure conditions. Whereas water quality is dependant on the flow at all preceding time steps irrespective of whether the flow is low or high. Also accurate prediction of the flow routing is particularly important for water quality analysis. Hence using a hydraulic model for predicting water quality in distribution requires a substantially different level of construction detail. This project would look at issues such as what degree of skeletonization is required, degree of calibration, use of tracer studies, and how to use these models in a realistic way.
- *Fate and characterization of natural organic matter (NOM) in distribution systems to develop water quality models:* NOM plays an important role in all distribution system water quality reactions. Yet for most systems NOM is only characterized as TOC. This does not give any indication of the behavior of NOM or its reactivity with disinfectants, biofilms, or discoloration. This project would examine what level of NOM characterization would be productive for modeling and understanding water quality in distribution.
- *Susceptibility of distribution systems to negative pressure transients:* Pressure transients (or surge) are waves of high and low pressure that pass through the water in pipes as a result of a sudden change in velocity. It is well established that pressure transients are a regular and common occurrence within many distribution systems. Although two AwwaRF funded projects; *Pathogen Intrusion into the Distribution System* and *Field Testing of Surge Modeling Predictions*, have established that pressure transients occur in drinking water systems and that potential sites of intrusion are often surrounded by contaminated water, there has been no research into the incidence and consequential risk of contamination of distribution pipes resulting from pressure transients. It is these two research needs that are to be addressed by this project.
- *Reference manual for evaluation of water quality coefficients:* Considerable effort has gone into the development of models of various water quality parameters. However, in order to apply these models, the modeler needs to be able to define values for the model coefficients. Guidance on and consistence of application will improve the comparability of models which is particularly important where models are being used as part of the regulatory process.
- *Impact of pipe wall factors on modeling water age and chlorine and chloramine decay:* The condition of the internal surfaces of pipelines from a modeling perspective is poorly understood. Many times the internal diameter of a pipe is incorrectly assumed. The project would develop guidelines for determining appropriate effective diameters and pipe wall factors for use in water age and disinfection residual modeling.
- *Internal pipeline scale characterization:* Nearly all chemical constituents that are modeled are greatly affected by the material and condition of the pipe wall. Where present, unlined cast-iron has a particularly strong impact on distribution system water

quality due to its rapid consumption of disinfectant residuals or due to the release of iron resulting in discolored water. Further development of models to predict water quality has been restricted by a lack of understanding on the nature of the scale which develops at the pipe wall. This scale will consist of corrosion products, hardness precipitates, and other deposits. Reactions cannot be modeled in detail without this knowledge. This project would analyze pipe scale under a wide variety of water conditions.

In addition to the six high priority projects suggested for funding, five additional research briefs were proposed:

- *Assessment of DBP Modeling Techniques for Compliance with Stage 2 Disinfectant and Disinfectant By-Product Rule:* The Stage 2 DBPR requires that utilities exercise better control on DBPs in water distribution systems. It is designed to reduce peak DBP concentrations using fixed sample sites and the Locational Running Annual Average (LRAA) for each site must comply with the MCLs. The IDSE proposes the use of modeling techniques to select these new monitoring locations, but the USEPA only provide limited guidance as to the types of models to be used, or how these models might be calibrated. It suggests the use of water quality distribution network models, but not mechanistic models. This project will assess and evaluate current water system and mechanistic modeling techniques for predicting Stage 2 DBPR and disinfectant residuals in drinking water distribution systems.
- *Characterization of Particles to Develop Models for Controlling Discoloration:* One of the most frequent customer complaints of utilities is related to discoloration events in the distribution system. The nature and composition of the particles involved in these events is not well defined or understood. As such an understanding of the characterization of discoloration and possible impact on the water quality is important and can be used to manage the risk of discoloration. Characterization of the chemical and biological nature of particles will give insight into the origin, initial source, and cause of mobilization and health implications. This information can be joined with hydraulic models to minimize discoloration events, and develop the proper methods to deal with the events, thereby minimizing the possible health effects on the customers.
- *Decision Tool for Bacterial Regrowth Control Options:* Bacterial biofilms in distribution systems contribute to a variety of compliance issues, including Total Coliform Rule (TCR) violations and elevated heterotrophic plate counts (HPC). It is also known that biofilms in distribution systems are reservoirs for EPA Drinking Water Microbial Contaminant Candidate List (CCL) organisms, including *Legionella*, *Aeromonas*, and *Mycobacterium spp*. Although the phenomena associated with the accumulation and persistence of these organisms is not completely understood, there has been knowledge accumulated in basic research and utility experience that provides insight on the relative importance of a variety of parameters contributing to the presence and activity of these organisms. The intent of this project is to bring together this information in a decision tool format, relying on the input of experts and available research, so that utilities can utilize the tool to perform a self-assessment and implement control strategies.

- *Modeling the Health Risks of Microbial Intrusion into Distribution Systems Operated with Chlorine and Chloramine Residuals:* Recent studies have demonstrated the susceptibility of distribution systems to intrusion during transient low or negative pressure events. Intrusion of contaminated water into the distribution system could have public health consequences depending on the amount of ingress and the ability of the distribution disinfectant residual to eliminate any pathogens present. The probability and extent of intrusion have been shown to be system specific. This project will include the development of a decision tool for assessing the options available to operators and asset planners for managing microbial water quality in relation to internal pipe regrowth and external intrusion.
- *Tracer Studies for Water Quality Model Calibration and Validation:* Traditional calibration and validation methods for water quality distribution network models include comparison of model results to field measurements of pressure and flow. An alternative (or additional) approach that has evolved over the past decade is the tracer study. Tracer studies have rarely been used, but when used they have been found to be especially effective as a calibration procedure for water quality studies. Consequently a sound approach needs to be developed for tracer studies to determine hydraulic characteristics that are required to support water quality model calibration and validation.

OTHER RESEARCH NEEDS

During the course of the workshop a number of further research needs were identified. While they were not the priority they still represent needs for future attention.

- *Disinfectant, DBP, Corrosion Product, and NOM Model:* Development of a model to describe interactions among disinfectants, DBPs, corrosion products, and NOM.
- *Nitrification Model:* Development of a model to predict the onset of nitrification
- *Operational Data Analysis:* Use of operational management data to analyze water quality problems. The key to this being the integrated use of different data sources including:
  - GIS
  - Customer and other business databases
  - Telemetry records
  - Simple models
  - Statistical models
- *In-Situ Test Distribution System:* Development of an in-situ test distribution system that can be used for various research projects. This might be based on an existing non-potable system (e.g. irrigation or industrial). Research performed on the test system might address:
  - Intrusion, surge, and potential contaminants
  - Chlorine or chloramine decay
  - Biofilms
  - Use of new tracer types
- *Risk-Based Assessment of Distribution System Operation:* This would bridge operational gaps and issues.

APPENDIX A
RESEARCH BRIEFS

Through a workshop attended by water utility personnel, academics, and consultants, eleven research briefs were drafted. These are intended to be reviewed by Awwa Research Foundation for possible issue as a 'Request for Research Proposal' (RFP). The workshop attendees identified six of the proposals as being of particularly high importance. The order within the lists does not indicate any priority.

Highest Priority Briefs

A.1 Guidelines for Developing, Calibrating, and Using Hydraulic Models for Water Quality Modeling
A.2 Fate and Characterization of Natural Organic Matter (NOM) in Distribution Systems to Develop Water Quality Models
A.3 Susceptibility of Distribution Systems to Pressure Transients
A.4 Reference Manual for Evaluating Water Quality Model Coefficients
A.5 Impact of Pipe Wall Factors on Modeling Water Age, Chlorine, and Monochloramine Decay
A.6 Internal Pipeline Scale Characterization

Other Briefs

A.7 Assessment of DBP Modeling Techniques for Compliance With Stage 2 Disinfectant and Disinfectant By-product Rule
A.8 Characterization of Particles to Develop Models for Controlling Discoloration
A.9 Decision Tool for Bacterial Regrowth Control Options
A.10 Modeling the Health Risks of Microbial Intrusion Into Distribution Systems Operated with Chlorine and Chloramine Residuals
A.11 Tracer Studies for Water Quality Model Calibration and Validation

A.1 GUIDELINES FOR DEVELOPING, CALIBRATING, AND USING HYDRAULIC MODELS FOR WATER QUALITY MODELING

OBJECTIVE

The objective of the research project is to produce detailed guidance to improve water quality modeling techniques to better support utility decision making. The project shall address the data and technical quality issues that are required to support water quality model calibration and validation including the use of tracer studies to determine hydraulic characteristics.

BACKGROUND

Hydraulic modeling has been utilized for a wide range of planning, design, and operational tasks. In recent years, it has also been used as a basis for more advanced water quality modeling. However, it has been recognized by the modeling profession that the level of detail required is generally greater when used to perform water quality modeling. The purpose of the project is to undertake research in order to develop guidance on what is required in a hydraulic model when used to support water quality modeling. In the absence of these guidelines, the results are too variable to allow cost-effective operational and capital improvement decision-making.

Traditional calibration and validation methods for water quality distribution network models include comparison of model results to field measurements of pressure and flow. An alternative, or additional, approach that has evolved over the past decade is the tracer study. Tracer studies have rarely been used, but when used they have been found to be especially effective as a calibration procedure for water quality studies. In a tracer study, a conservative chemical (fluoride, calcium, etc) is injected at a point in the distribution system and its movement through the distribution system is measured. The field results are then compared to model results and adjustments made in the model parameters to better reflect field results. However, at this point there is no standard protocol for performing a tracer study in the distribution system. This project will draw upon past AwwaRF research projects on conducting tracer studies in clear wells and distribution system storage facilities to develop a cost-effective distribution system tracer method.

RESEARCH APPROACH

The general approach to the project will encompass an assessment of past research, modeling investigations, case studies, and an open forum workshop comprising industry experts and specialists. The specific tasks to be completed shall include, but not be limited to:

- The need and data requirements for Extended Period Simulation (EPS) models required for water quality models
- Determination of the impact of skeletonization within water quality models. Comparison of the results between all pipe and skeletal models.
- Provision of guidance on skeletonization processes
- Minimum level of detail for hydraulic calibration for models used for water quality modeling

- Trace chemical selection, dose, and injection methods
  - Operational and regulatory issues
  - Protocols (frequency, location, methodology, validation)
  - Determination of frequency of tracer studies to analyze supply and demand characteristics and allocation (e.g. seasonal change, source change)
  - Instrumentation available to ensure successful input and validation data
- Manual or automated methods for adjusting hydraulic parameters to reflect tracer study results
- Guidance on the appropriate level of model detail and calibration of water quality models in relation to the specific water quality problem
- Appropriate measurement and credibility of output for the problem
- Confidence levels for input and calibration data and results
- Validation of existing models (opportunity to expand for problem area)
- Minimum range of instrumentation for water quality modeling
- Model maintenance
- Surrogate indicators

The output of this research shall be a comprehensive guidance on distribution system hydraulic and water quality modeling for large and small utilities each with different levels of available data and resources. The documentation will include protocols for using tracer studies to enhance model calibration and provide supporting information for determining confidence limits for model calibration and predictive analysis results.

BUDGET AND TIME SCHEDULE

The anticipated project budget is $400,000 ($300,000 for the model guidance manual and $100,000 for the tracer studies element of the task). The project period should be realistic, anticipate possible starting delays, and provide ample time for the writing of final reports and for Project Advisory Committee (PAC) review of project results. The project program is 36 months.

A.2 FATE AND CHARACTERIZATION OF NATURAL ORGANIC MATTER (NOM) IN DISTRIBUTION SYSTEMS TO DEVELOP WATER QUALITY MODELS

OBJECTIVE

This project will develop methods that enable water utilities to characterize NOM for predicting water quality changes in distribution systems.

BACKGROUND

NOM plays a key role in disinfectant loss, formation of DBPs, corrosion, and production of biofilms. In order to enable these processes to be modeled effectively, there is a fundamental need for a method of characterizing NOM which is relevant to the wide range of distribution system water quality problems which it influences. Currently, no practical method for characterizing NOM is available. The characterization would fractionate NOM into specific subgroups based on reactivity with bacterial regrowth, disinfectant loss, DBPs, and corrosion. With this information water utilities could alter treatment to remove NOM to improve distribution system water quality.

RESEARCH APPROACH

The research approach should include the following elements:

- An initial review to determine and understand the potential impacts of NOM on distribution system water quality problems (e.g. biofilms, disinfectant loss, DBPs, and corrosion)
- A series of experiments based on the initial review to determine what paradigm of NOM fractionation would be appropriate as an input to water quality models and be a viable tool for controlling distribution system water quality problems. This should center on NOM fractions that can be easily isolated (e.g., charge, hydrophilic, hydrophobic) and can be removed by advanced treatment process.
- Pilot studies with a model system to validate the initial experiment plan. This should encompass a system that can replicate distribution system processes and in particular surfaces.
- If practicable, further validation should be implemented on full-scale systems The final report should contain guidance to enable water systems to control NOM to improve distribution system water quality.

BUDGET AND TIME SCHEDULE

The anticipated project budget is $350,000. The project period should be realistic, anticipate possible starting delays, and provide ample time for the writing of final reports and for Project Advisory Committee (PAC) review of project results. The project program is 36 months.

A.3 SUSCEPTIBILITY OF DISTRIBUTION SYSTEMS TO PRESSURE TRANSIENTS

OBJECTIVE

The objective of the research project is to determine the factors which control the susceptibility of distribution systems to deterioration in water quality as a consequence of pressure transients and particularly the resultant intrusion of contaminated water during a transient event. The project report will give utilities an understanding of pressure transients and their implications for water quality and provide guidance on the monitoring and control of pressure transients.

BACKGROUND

Pressure transients (or surge) are waves of high and low pressure that pass through the water in pipes as a result of a sudden change in velocity. It is well established that pressure transients are a regular and common occurrence within many distribution systems. Pressure transients can arise from normal operational activities, including pump switching, valve operation, and mains flushing. They can also be caused by uncontrolled operations including main breaks, sudden changes in demand (including the actions of a large single consumer), uncontrolled pump start up or stoppage (commonly caused by power failures), air valve slam, and operation of fire hydrants.

Negative pressure resulting from a transient event gives the potential for intrusion to occur at pipe joints, breaks in the pipe wall, and through poorly maintained or defective network apparatus. Any point of potential intrusion represents a risk to water quality.

Most of the currently available distribution system hydraulic analysis software will only model steady state hydraulic conditions; extended period simulation (EPS) being a repeating sequential steady state analysis at a defined time step. A limited number of network analysis software packages currently incorporate a transient model, usually with a network size analysis limitation. There are also a number of transient analysis tools for pipe systems which could be used to analyze transients in distribution networks. However, transient analysis software generally does not account for continuous changes in supply and demand as found in the periphery and smaller pipes of systems, but generally masked in the heart of a network.

Previous research focused on documenting the frequency and magnitude of pressure transient events using high-speed, electronic, pressure data loggers. The logged results from a limited sample of systems identified a number of negative pressure transient events which were mostly related to pump-shutdowns. Public health concern of negative pressure transients was heightened by studies that documented the presence of human enteric viruses in 56% of the soil and water samples adjacent and exterior to distribution system pipelines. Limited research to date has linked field sampling data at a point of potential intrusion with recorded microbial noncompliance data.

Although two AwwaRF funded projects; *Pathogen Intrusion into the Distribution System* and *Field Testing of Surge Modeling Predictions*, have established that pressure transients occur in drinking water systems, and that potential sites of intrusion are often surrounded by contaminated water, there has been no research into the incidence and consequential risk of contamination of distribution pipes resulting from pressure transients. It is these two research needs which are to be addressed by this project.

RESEARCH APPROACH

The project is intended to address the following three questions;

1. What are the characteristics of the causes of pressure transients?
2. How significant are pressure transients in different systems?
3. How significant is intrusion in terms of water quality taking into account:
   - Compliance sampling
   - Microbial failures
   - Seeding for regrowth
   - Propagation through the system
   - Customer contact

The accuracy and credibility of distribution network models is currently constrained by the availability and quality of the base data used to construct and calibrate models. The quality and volumes of data needed to evaluate transient conditions within distribution are significantly greater. The researchers should address this concern so that questions are not raised about the credibility of the analyses used in the project.

The research approach should include the following elements:

- Develop transient analysis models for a representative number of different networks with a comprehensive range of causal characteristics. The models can be developments of calibrated hydraulic models built for other purposes (e.g. for System Specific Studies (SSS) as part of an Initial Distribution System Evaluation (IDSE). The transient models may include additional data not necessarily included in either steady state or EPS hydraulic models (e.g. pump and valve characteristics, pump and valve performance curves, pump speed and inertia data, and operating regime).
- Establish the principle controlled and uncontrolled activities resulting in transient pressures and the risk of causing a significant event. For the high risk causes, carry out sensitivity analyses to determine the significance of model inputs and acceptable levels of model simplification.
- Simulate pressure transient events to characterize the response of different types of network. A common set of simulated transient events should be used for each model, including at least:
  - Complete loss of pumping (e.g. power outage)
  - A major main break in a key trunk line
  - Hydrant flushing in a vulnerable area (low elevation)
  - Rapid valve closure of a major distribution main
  - Large consumer demand change at network extremity
- For each event, results will be evaluated to establish the magnitude and duration of the pressure transient, the number of nodes exhibiting negative pressures and the total duration of the negative event and hence the hydraulic characteristic conditions likely to promote a transient event
- The amount of water entering the system will be estimated using a uniform set of parameters (e.g., orifice size, external head, leakage rate, etc.) to allow comparisons

between systems, including categorizing systems into low, medium, and high vulnerability to intrusion

- Produce a risk assessment comprising all the factors which affect the risk of intrusion and the consequential impact on failure of compliance samples and public health. An assessment shall be made of potential overall risk and the investigations should establish if a set of parameters can be used to classify systems at high risk of intrusion. If possible, data from WaterStats will be accessed to try to extrapolate the impact of this analysis to the water industry.

BUDGET AND TIME SCHEDULE

The anticipated project budget is $350,000. The project period should be realistic, anticipate possible starting delays, and provide ample time for the writing of final reports and for Project Advisory Committee (PAC) review of project results. The project program is 36 months.

A.4 REFERENCE MANUAL FOR EVALUATING WATER QUALITY MODEL COEFFICIENTS

OBJECTIVE

To produce a manual which gives guidance on evaluating the coefficients for all water quality models which are at a sufficient stage of development for utility application. The evaluation methods would include standard values, approximation methods, and measurement methods.

BACKGROUND

Considerable effort has gone into the development of models of various water quality parameters. However, in order to apply these models, the modeler needs to be able to define values for the model coefficients. Some models, such as the widely publicized SANCHO model, contain over 20 different input variables or coefficients. Published values for chlorine and chloramine decay coefficients have varied in magnitude by 10^3-fold or more. It is often difficult for a modeler to obtain values for these coefficients and there is a danger that values will be assumed to be the same as in previous application of the model where different conditions apply. The modeler therefore needs guidance on when and where to use site specific derived values as against general guidance figures. There is also benefit in standardizing evaluation methods for coefficients. Guidance on and consistence of application will improve the comparability of models which is particularly important where models are being used as part of the regulatory process.

RESEARCH APPROACH

The research project will be predominantly a literature review and review of current practice. The following steps would be involved:

1. Identify the current range of water quality models
2. Identify the variables and coefficients in each model, including indicating:
   - The sensitivity of the model results to the accuracy of each parameter and
   - Any significant relationship between parameters that will impact the model sensitivity
3. Undertake a literature review to:
   - Obtain standard values, typical values, and the range of values for each coefficient
   - Discuss the confidence in those standard values
   - Identify whether the values are globally applicable or site specific
4. Where global values are not appropriate, give methods for measurement, such as standard methods for jar tests

The study should include, but not be limited to:

- Chlorine models (zero, first, and second order)
- Chloramine models

- THM growth
- Typical microbial coefficients including all coefficients in the SANCHO model
- Sediment transport models
- Diffusion and mass transport coefficients
- Thermal effects (Arrhenius coefficients)

The final report shall be structured as a reference manual which explains the purpose of each coefficient, indicates the sensitivity of model output to each variable, identifies whether the value is dependant on specific factors (e.g. pipe material), gives typical values for the parameter, and defines methods for evaluating it.

BUDGET AND TIME SCHEDULE

The anticipated project budget is $200,000. The project period should be realistic, anticipate possible starting delays, and provide ample time for the writing of final reports and for Project Advisory Committee (PAC) review of project results. The project program is 18 months.

A.5 IMPACT OF PIPE WALL FACTORS ON MODELING WATER AGE, CHLORINE, AND MONOCHLORAMINE DECAY

OBJECTIVE

The objective of the study is to investigate the influence of the pipe wall on water age and disinfection decay and thereby develop guidance for predictive modeling. The detailed objectives include:

- To characterize pipe wall factors, particularly actual internal diameters, pipe wall roughness coefficients, and wall decay coefficients for characteristic pipe types
- To investigate the mechanisms which control chlorine consumption at the pipe wall
- To validate current modeling methods and understand their limitations

BACKGROUND

Age of water within a network is increasingly estimated either as a surrogate for or as part of the analysis of another water quality parameter. However age calculations are often performed using nominal pipe diameters instead of the actual internal diameter. The actual diameter may be significantly less than the nominal diameter due to tuberculation and other cross-sectional area reduction factors. The consequence of using nominal diameters is that water age is overestimated and any subsequent predictions may therefore be based on over-estimations of chemical and biological rates of change. It is widely acknowledged that chlorine decay is attributable to both bulk water processes and interaction with the pipe wall. Generally, bulk water processes are relatively well understood and can often be modeled with reasonable accuracy. However, the interaction with the pipe wall, including issues such as corrosion, biofilms, and oxidation processes, are generally more complex and difficult to model.

The project should build on the work published in AwwaRF Project *Characterization and Modeling of Chlorine Decay in Distribution Systems* (Vasconcelos et al. 1996). This project included research into the effect of different pipe materials on wall decay and contributed to the development of the EPANET model.

More recently, some anomalies with current modeling approaches have been noted by several researchers:

- The rate of decay has been observed to increase with increased velocity
- Attempts to characterize values of decay coefficients for different pipe materials have resulted in an extremely wide range of values varying by orders of magnitude for apparently similar pipe materials
- In some unlined cast-iron pipes, the wall decay has been found to fit zero order decay more closely than first order decay

The chlorine decay model used within the EPANET network modeling software has been widely adopted in other software. This model incorporates a mass transfer model for the radial transfer of chlorine from the bulk flow to the pipe wall. The model was derived from process plant theory and its application to distribution networks has not been fully validated. Although many successful chlorine decay models have been built using this model, it is unclear whether the mass transfer element contributed to that success. In particular there is a length term which is difficult to apply in network models since pipe length is dictated by the number and location of nodes which to a large extent is an arbitrary decision.

RESEARCH APPROACH

The study should cover both chlorine and monochloramine decay, but must clearly differentiate between the two in all reports. These research should include the following specific studies:

- Review the implications of using nominal pipe diameters as opposed to actual pipe diameters. This should not only consider the implications for water age but also the implications for radial mass transfer, surface area contact, and wall decay.
- Derive a method of determining average effective diameter and its impact on roughness (Hazen-Williams; Darcy-Weisbach) and travel time over a representative pipe reach
- Undertake a thorough literature review of existing research into wall decay mechanisms, highlighting where there is consensus or conflict in conclusions. This review should cover the source material for the EPANET wall decay model and assess whether there are any limitations in the way that it has been derived.
- Develop protocols and methodologies for deriving the values of chlorine and monochloramine wall decay coefficients over a representative pipe reach
- Hence, through a series of field studies:
  - Give guidance values for decay coefficients representative of different relevant pipe materials, age, and water types
  - Validate the EPANET wall decay model confirming whether any key variables such as velocity affect chlorine decay in a manner not currently accounted for. This will require structuring the field trials so that the various components of the model are tested effectively. Modifications or simplifications to the model should be recommended and tested where appropriate.

BUDGET AND TIME SCHEDULE

The anticipated project budget is $300,000. The project period should be realistic, anticipate possible starting delays, and provide ample time for the writing of final reports and for Project Advisory Committee (PAC) review of project results. The project program is 36 months.

A.6 INTERNAL PIPELINE SCALE CHARACTERIZATION

OBJECTIVE

With ageing, pipe walls become lined by scale. This scale will consist of corrosion products, hardness precipitates, and other deposits. The nature and composition of this scale will have a strong influence on water quality. The objective of this project is to characterize the structure, composition, and mineralogy of corrosion scales and deposits removed from the surfaces of various types of pipes collected from a number of drinking water distribution systems.

BACKGROUND

Corrosion scales and deposits can have a major impact upon water quality by:

- Acting as disinfectant sinks
- Providing environments that support biofilm activity, thereby concentrating organic material and contaminants such as arsenic
- Controlling the rate and extent of metal release

The structure (physical geometry) of scales will control the roughness and internal diameter of pipes and so can have a significant impact on the hydraulic behavior of a network.

Critical water quality judgments in distribution systems are often based on pipe materials. However, it is the material at the interface between the pipe and the water which is of fundamental importance in terms of water quality reactions. This material will often be a corrosion scale, but there is little information on chemical composition of these scales.

Improved knowledge on scale composition would enable utilities to understand how water chemistry interacts with the process of scale formation and the resultant water quality problems. This would enable treatment processes to be optimized to control scale-related water quality issues most effectively. A deeper understanding of scale characteristics and behavior will lead to improved modeling of chlorine decay, biofilm formation, DBP growth, and corrosion problems.

RESEARCH APPROACH

The approach to the project will revolve around characterizing corrosion scales and deposits removed from pipes collected from a number of water utilities. Water utilities will be selected so as to represent a range of water chemistries with due regard for oxidant type and the use of corrosion inhibitors. Samples shall be collected from a range of pipe materials including iron, copper, lead, and brass. Corrosion scales removed from the pipe sections shall be visually described and analyzed using a variety of approaches including, but not limited to, x-ray diffraction (mineralogy), digestion or ICP-MS (elemental composition), SEM-EDXA (imaging or elemental composition) and mossbauer. The nature of the scales shall be compared in relation to:

- Water chemistry
- Disinfectant losses
- DBP formation

- Frequency of colored water complaints in the distribution system
- Records of lead, copper, and iron concentrations

The final report should:

- Characterize a range of different corrosion scales and attempt to explain the cause of observed differences between samples
- Review the implications of the findings for modeling pipe wall water quality interactions, highlighting where existing models are inadequate, and suggesting improvement
- Develop guidance on how to use scale information to improve treatment

BUDGET AND TIME SCHEDULE

The anticipated project budget is $400,000. The project period should be realistic, anticipate possible starting delays, and provide ample time for the writing of final reports and for Project Advisory Committee (PAC) review of project results. The project program is 36 months.

A.7 ASSESSMENT OF DBP MODELING TECHNIQUES FOR COMPLIANCE WITH STAGE 2 DISINFECTANT AND DISINFECTANT BY-PRODUCT RULE

OBJECTIVE

The objective of the proposed work is to provide guidance on how to meet the requirements of the Stage 2 Disinfectant and Disinfectant By-Product Rule (DBPR) by modeling and to make recommendations on further research that should be conducted to overcome any limitations to the application of these models.

BACKGROUND

Trihalomethanes (THMs) and haloacetic acids (HAAs) are regulated in the USA by the DBPR. The MCLs are 80 μg/L for THMs and 60 μg/L for HAAs. Fixed sample sites are used and the Locational Running Annual Average (LRAA) for each site must comply with the MCLs. An Initial Distribution System Evaluation (IDSE) is under consideration. If implemented, the IDSE will require utilities to select representative sample sites, unless the water utility can demonstrate from two years of data that distribution samples are less than 40 μg/L for THMs and 30 μg/L for HAAs. For most systems the IDSE will result in the need for new monitoring points to more accurately represent typical high concentrations of total trihalomethanes (TTHM) and HAAs.

The DBPR proposes the use of modeling techniques as part of the IDSE to select these new monitoring locations, but the USEPA only provides limited guidance as to the types of models to be used, or how these models might be calibrated. It suggests the use of water quality distribution network models, but not mechanistic models. However, both types of models have applicability to this problem, especially in combination.

A water quality distribution network model is a computer program that simulates the hydraulic and water quality behavior of the water in a distribution system. Hydraulic models can be used to predict water age and can track the movement from each plant or supply point within the system. It can be used to provide an understanding of the influence zone of each system entry point and identify the blending zones. Another aspect of modeling for the DBPR is to be able to mechanistically characterize the changes that occur in the concentrations of the various DBP species in a distribution network. These changes are influenced by the choice of primary disinfectant (chlorine or chloramines), the physical and chemical characteristics of the water entering the systems, and by the pipe material in the network. For example, it is well known that TTHM concentrations are a function of chlorine concentration, temperature, pH, concentration of Natural Organic Matter (NOM), time of formation, and the concentration of inorganic constituents in the water.

RESEARCH APPROACH

An evaluation will be carried out of the current modeling techniques (both water quality distribution network and mechanistic) used for predicting Stage 2 DBPs and disinfectant residuals in drinking water distribution systems. The research will combine both these models into a computer program for predicting Stage 2 DBP levels and chlorine residuals at key control points in the network.

Recommendations will be made on what further research should be conducted to overcome any limitations on the application of these models.

SCHEDULE AND BUDGET

Due to the immediacy of these standards coming into force this research should be carried out as rapidly as possible for the results to be of full benefit to the companies affected. Thus, a project period of 12 months and a budget of $150,000 are anticipated.

A.8 CHARACTERIZATION OF PARTICLES TO DEVELOP MODELS FOR CONTROLLING DISCOLORATION

OBJECTIVE

The objective of this project is to determine and characterize the composition of particles causing discoloration events in distribution systems and how this can be used to manage the risk of discoloration. Guidance will be provided on how this information can be joined with hydraulic models to minimize discoloration events and how proper methods can be developed to deal with the events, thereby minimizing the possible health effects on the customers.

BACKGROUND

One of the most frequent customer complaints of utilities is related to discoloration events in the distribution system. The nature and composition of the particles involved in these events is not well defined or understood. For example, recent studies have indicated that low levels of arsenic in the finished water entering the distribution system can concentrate with iron deposits and be released during discoloration events. Additionally, it is well know that bacteria can survive and multiply in the biofilm.

Characterization of the chemical and biological nature of particles will give insight into their origin, initial source, cause of mobilization, and health implications.

RESEARCH APPROACH

The research approach will include, but not be limited to:

- A literature review of available information
- Collection of samples from various utilities representing a variety of source waters and water quality characteristics, during and after discoloration events, and during controlled flushing events. The samples collected after the discoloration events will be used to determine when the water quality has returned either to the same quality as prior to the event or to a point where the water quality is no longer being impacted. The flushing velocity used to collect samples will be recorded and analyzed so as to assist with the development of flushing models.
- Development of an analytical matrix to characterize the physical, chemical, and microbiological nature of the particles in a sample, including potential health-related parameters
- Analysis of the samples using this analytical matrix
- A spatial analysis to identify the characteristics of different typical parts of the network such as trunk mains, dead ends, and proximity to zone valves
- Development of a risk matrix showing the spatial distribution of the risk for each distribution system
- Application of the information from this risk matrix to the hydraulic model to assist in defining the risk areas in the distribution system

BUDGET AND TIME SCHEDULE

The anticipated project budget is $400,000. The project period should be realistic, anticipate possible starting delays, and provide ample time for the writing of final reports and for Project Advisory Committee (PAC) review of project results. The project program is 36 months.

A.9 DECISION TOOL FOR BACTERIAL REGROWTH CONTROL OPTIONS

OBJECTIVE

The objective of the proposed work is to provide utilities with a decision tool to enable them to respond to bacterial regrowth and biofilm problems. The intention is to give utilities the information needed to form a biofilm management system, therefore allowing them to perform a self-assessment and implement control strategies so that they can become compliant with violations of the total coliform rule (TCR), to control elevated heterotrophic plate counts (HPC), and to address the mitigation of EPA Drinking Water Microbial Contaminant Candidate List (CCL) organisms such as Legionella or Mycobacterium. The tool will also determine areas of uncertainty in the ability to control these organisms. After use of the tool, a utility should be able to determine the relative vulnerability of their system to the impact of these problems.

BACKGROUND

Bacterial biofilms in distribution systems contribute to a variety of compliance issues, including TCR violations and elevated HPC counts. It is also known that biofilms in distribution systems are reservoirs for CCL organisms, including Legionella, Aeromonas, and Mycobacterium spp. Although the phenomena associated with the accumulation and persistence of these organisms is not completely understood, there has been knowledge accumulated in basic research and utility experience that provides insight into the relative importance of a variety of parameters contributing to the presence and activity of these organisms.

RESEARCH APPROACH

A compilation of the available information on factors leading to the presence of key organisms and HPC in distribution systems will be made. It is preferred that experts in the field then be consulted to obtain input on the relative importance of these factors The experience of several utility case studies will be included to provide practical examples of applied principles. The data will be compiled into an interactive (potentially web-based) program that utilities can use as a self-assessment tool. The tool will have the ability to be updated as new expert information is obtained and be user-friendly with adequate supporting information. An analysis of critical gaps in the knowledge base will be conducted and recommendations for future research provided.

BUDGET AND TIME SCHEDULE

The anticipated project budget is $325,000. The project period should be realistic, anticipate possible starting delays, and provide ample time for the writing of final reports and for Project Advisory Committee (PAC) review of project results. The project program is 36 months.

A.10 MODELING THE HEALTH RISKS OF MICROBIAL INTRUSION INTO DISTRIBUTION SYSTEMS OPERATED WITH CHLORINE AND CHLORAMINE RESIDUALS

OBJECTIVE

To develop a decision tool for assessing the options available to operators and asset planners for managing microbial water quality in relation to internal pipe regrowth and external intrusion.

BACKGROUND

Recent studies have demonstrated the susceptibility of distribution systems to intrusion during transient low or negative pressure events. Intrusion of contaminated water into the distribution system could have public health consequences depending on the amount of ingress and the ability of the distribution disinfectant residual to eliminate any pathogens present. The probability and extent of intrusion has been shown to be system specific.

Many utilities have adopted, or are considering adopting, chloramines as their disinfectant residual to reduce the formation of DBPs in distribution to meet new regulations. While chloramines are more effective than chlorine for biofilm control, they have higher 'CT values' (product of concentration and contact time) particularly for viruses.

The significance of contravening standards due to regrowth or due to intrusion is not well understood. Utilities need to understand the risks and the options available so as to minimize or balance the health risks against the regrowth risks.

RESEARCH APPROACH

This research project will require expertise in both hydraulic modeling and practical microbiology.

The researcher will run various simulations of transient intrusion into different distribution networks using hydraulic models with surge capability. Under each simulation, the dispersion of characteristic organisms in the network and the effects of disinfectant residuals (chlorine and chloramines) will be determined. Characteristic organisms with robust CT will be selected (e.g. Enteroviruses, Legionella). The findings from the hydraulic modeling simulations should be validated under laboratory or pilot scale conditions through selective experiments.

The researchers are also required to evaluate the extent of the risk from intrusion and the options of control with residual disinfectant, increasing pressures, provision of surge vessels, and improved leakage control (rehabilitation, replacement etc).

The final report will identify the fate of the model pathogens entering the distribution system through intrusion and options for minimizing health risks in distribution networks operating with chlorine and chloramine residuals.

BUDGET AND TIME SCHEDULE

The anticipated project budget is $200,000. The project period should be realistic, anticipate possible starting delays, and provide ample time for the writing of final reports and for Project Advisory Committee (PAC) review of project results. The project program is 24 months.

A.11 TRACER STUDIES FOR WATER QUALITY MODEL CALIBRATION AND VALIDATION

OBJECTIVE

The objective is to develop an approach for tracer studies to determine hydraulic characteristics that are required to support water quality model calibration and validation

BACKGROUND

Traditional calibration and validation methods for water quality distribution network models include comparison of model results with field measurements of pressure and flow. An alternative (or additional) approach that has evolved over the past decade is the tracer study. Tracer studies have rarely been used, but when used they have been found to be especially effective as a calibration procedure for water quality studies. In a tracer study, a conservative chemical (fluoride, calcium, etc) is injected at a point in the distribution system and its movement through the distribution system is measured. The field results are then compared to model results and adjustments made in the model parameters to better reflect field results. However, at this point there is no standard protocol for performing a tracer study in the distribution system. This project will draw upon past AwwaRF research projects on conducting tracer studies in clear wells and distribution system storage facilities to develop a cost effective distribution system tracer method.

RESEARCH APPROACH

The general approach to the project will encompass an assessment of past research, modeling investigations, case studies, and an open forum workshop comprising industry experts and specialists.

The specific tasks to be completed shall include, but not be limited to:

- Trace chemical selection, dose and injection methods
- Operational and regulatory issues
- Protocols (frequency, location, methodology, validation)
- Determination of frequency of tracer studies to analyze supply and demand characteristics and allocation (e.g. seasonal change, source change)
- Instrumentation available to ensure successful input and validation data
- Manual or automated methods for adjusting hydraulic parameters to reflect tracer study results

BUDGET AND TIME SCHEDULE

The anticipated project budget is $300,000. The project period should be realistic, anticipate possible starting delays, and provide ample time for the writing of final reports and for Project Advisory Committee (PAC) review of project results. The project program is 24 months.

APPENDIX B
GENERAL MODELLING METHODS

This Appendix gives details about two modeling methods which can be used as the basis for water quality models. The chapter focuses on the detail of the methods such as the mathematical aspects rather than their application to specific water quality modeling scenarios which is the focus in the main body of the report. These details are presented as a separate appendix because they are relevant to a broad range of different water quality issues and so do not fit into the structure of the preceding report.

Computational Fluid Dynamics (CFD)

CFD is generally taken to mean the numerical modeling of the motion of fluids based on the application of basic physical principles. Conventional modeling techniques such as the models of pipe flow incorporated into network models only predict the motion in one dimension i.e. axially to the pipe. CFD can model the behavior of fluid in three dimensions. This enables the modeling of eddies and mixing patterns which are of great significance to water quality. CFD generally involves defining a three dimensional mesh which splits the fluid into a large number of small elements. The behavior of these elements is then predicted using the Navier-Stokes set of simultaneous differential equations describing fluid flow. There are five equations covering:

- Conservation of mass
- Conservation of momentum (three equations, one for each dimension)
- Conservation of energy

In addition to the above, CFD analysis can also model chemical reactions, heat transfer, multi-phase flow (water, gases, and particles), and their interaction, all of which are desirable for detailed water quality modeling.

There are numerous commercially available CFD modeling packages. The software was originally developed for the aerospace and nuclear industries, but is now suitable for a wide range of applications. Hence, most packages are general purpose and not specific to the water industry. CFD is not currently a pragmatic method of modeling an entire distribution network as the computing power required would be excessive. Furthermore, the predictive capability would be no better than that of conventional network models due to the inherent uncertainty in pipe roughness and other unmeasured network properties. The primary application for CFD in the water industry is to model the behavior of storage tanks.

Benefits and Limitations of CFD

The benefits of CFD modeling include:

- It is the only pragmatic method of analyzing flow paths within complex structures
- It is suitable for the analysis of common inlet and outlet or standpipe tank designs and where there is poor mixing within the tank

- The software enables the modeling of complex interactions between water quality parameters, providing the behavior can be defined by numerical rules
- Many of the more detailed software packages have sophisticated graphical methods for viewing results which allow users to gain a better understanding of the three dimensional behavior within the tank
- Since the models are primarily based on fundamental physical principles, they are more reliable than more empirical models if the boundary conditions are significantly altered
- The models allow dynamic inertial effects to be predicted which can be significant in tanks. If a tank has been standing for some time with no flow in or out, the flow path upon restarting flows can be significantly different from the flow paths that would occur if it was operated with a steady through flow.

The limitations of CFD modeling include:

- High cost of software
- The need for skilled modelers
- It can be time consuming to model complex geometries
- Model run times can be high, restricting the ability to analyze time-varying demands or multiple scenarios. As a result studies are often based on a restricted number of steady state analyses.

Artificial Neural Networks

An artificial neural network is an empirical model which learns from representative historical data, rather than being programmed with specific laws. In so doing, it can discover complex non-linear relationships between input and output variables, accommodating subtle intricacies that are overlooked when the problem is reduced to relatively simple mathematical functions. To date ANNs have not been widely applied to water quality modeling, and so they do not feature strongly in this report. However their potential benefits have raised considerable interest and so ANNs are briefly reviewed here.

Water companies increasingly collect vast quantities of data about the behavior of their networks for operational and regulatory purposes, much of this data is likely to be related, but that relationship may be hard to define. It has been argued that ANNs are likely to be well suited to applications within water treatment and distribution (Rodriguez and Sérodes 1994) and have been applied to the modeling of free chlorine and THMs (Rodriguez et al. 1997; Rodriguez, Milot and Sérodes 2003). While programming the models with data is in principle straightforward, a considerable level of expertise is required to control the fundamental structure of the model. Unless software packages are developed which include predefined model structures for application to water distribution problems, they are unlikely to become a routine tool. While ANNs are capable of identifying complex relationships, they may struggle where there is a high degree of inaccurate or redundant data. Hence ANNs are probably most suited to process control rather than generalized analysis of distribution systems. In particular they are likely to struggle when applied to microbial water quality where relationships with other parameters result in an increased risk rather than a pre-determinable value. They may, however, be useful for coefficient estimation in real time or near time modeling.

Genetic Algorithms

Genetic algorithms (GA) are multi-parameter optimization methods which simulate the process of natural selection to converge upon improved solutions. The process is particularly effective for problems which involve large sets of data to be optimized. This is the case with many water distribution optimization problems where the performance of the system is controlled by the combined effect of all the individual components within the system. Various methods have been developed, but they generally involve the following processes:

1. An initial set or 'population' of different possible solutions are generated
2. In common with other optimization methods, an 'objective function' is defined, such as total cost. The objective function is used as the means of assessing the performance of the solution, the aim being to maximize (or minimize) the objective function.
3. The GA computes a 'fitness value' based on the objective function
4. Pairs of solutions are selected for 'reproduction.' The greater the fitness, the more likely that a solution will be selected for reproduction. Components of the solution are exchanged (crossover) between the two mating solutions to produce new hybrid solutions. Some methods induce occasional 'mutation' in this crossover process. This mutation process prevents the optimization stagnating due to identical parents.
5. Steps 3 and 4 are repeated until the objective function identifies a satisfactory solution

Various applications for GAs include:

- Scheduling of pumps to minimize operational costs
- Auto-calibration of network models (see Chapter 4)
- Identifying optimal sampling locations (Kapelan, Savic and Walters, 2003; Vítovský et al. 2003)
- Identifying the most cost effective renovation method (Halhal et al. 1997, Wu et al. 2001)
- Optimal sizing of new pipes or storage (Strafaci 2001, Boulos et al. 2000)
- Optimal dosage for source and booster chlorination (Munvalli and Mohan Kumar 2003)
- Network operation (setup) to optimize water quality (Brdys et al. 1995)

REFERENCES

Ackers, J., M. Brandt, and J. Powell. 2001. *Hydraulic Characterisation of Deposits and Review of Sediment Modelling*. London, England: United Kingdom Water Industry Research.

Ackers, J., D. Butler, and R. May. 1996. *Design of Sewers to Control Sediment Problems*. CIRIA Report 141. London, England: CIRIA.

Ackers, P., and W.R. White. 1973. Sediment Transport: New Approach and Analysis. *Journal of Hydraulics Division, ASCE*, 99(HY11):2041–2060.

Alegre, H., and S.T. Coelho. 1993. A Methodology for the Characterization of Water Consumption. In *Integrated Computer Applications in Water Supply*. Vol. 1. Edited by B. Coulbeck. Taunton, England: Research Studies Press.

Amy, G.L., P.A. Chadik, and Z.K. Chowdhury. 1987. Developing Models for Predicting Trihalomethane Formation Potential and Kinetics. *Journal of the American Water Works Association*, 79(7):89–97.

Axworthy, D.H., and B.W. Karney. 1996. Modeling Low Velocity/High Dispersion Flow in Water Distribution Systems. *Journal of Water Resources Planning and Management*, 122(3):214–217.

Berghout, B.L., and G. Kuczera. 1997. Network Linear Programming as Pipe Network Hydraulic Analysis Tool. *Journal of Hydraulic Engineering*, 123(6):549–559.

Black & Veatch. 2000. *Distribution System Water Quality Changes Following Implementation of Corrosion Control Strategies*. Denver, Colo.: AwwaRF and AWWA.

Boulos, P.F., T. Altman, P.A. Jarrige, and F. Collevati. 1994. An Event Driven Method for Modelling Contaminant Propagation in Water Networks. *Applied Mathematical Modelling*, 18:84–92.

Boulos, P.F., T. Altman, P.A. Jarrige, and F. Collevati. 1995. Discrete Simulation Approach for Network-Water-Quality Models. *Journal of Water Resources Planning and Management*, 121(1):49–60.

Boulos, P.F., Z.Y. Wu, C.H. Orr, and J.J. Ro. 2000. Least Cost Design and Rehabilitation of Water Distribution Systems Using Genetic Algorithms. In *Proc. of the AWWA Information Management and Technology Conference*. Denver, Colo.: AWWA.

Brandt, M. 2000. Taste Action Plan, Customer Perception of Drinking Water Taste. Presented at CIWEM Metropolitan Branch Meeting, London, January 26, 2000.

Brandt, M., J. Creasey, and R. Perkins. 1999. Making Decisions That Matter With Water Quality Models. In *Proc. of the International Conference on Computing and Control for the Water Industry*. Hertfordshire, England: Research Studies Press.

Brdys, M.A., H. Puta, E. Arnold, K. Chen, and S. Hopfgarten. 1995. Operational Control of Integrated Quality and Quantity in Water Systems. In *Proc. of International Federation of Automatic Control Symposium on Large Scale Systems: Theory and Applications*. Oxford: Pergamon Press.

Brendan, B.L., and G. Kuczera. 1997. Network Linear Programming as Pipe Network Hydraulic Analysis Tool. *Journal of Hydraulic Engineering*, 123(6):549–559.

Buchberger, S.J., J.T. Carter, Y. Lee, and T.G. Schade. 2003. *Random Demands, Travel Times, and Water Quality in Deadends*. Denver, Colo.: AwwaRF and AWWA.

Burgess, P.L., D.C. Vitanage, A. Holmes, and P.J. Bliss. 1996. Determination of the Importance of Wall Effects in Modelling Chlorine Residuals Within Distribution Systems. In *Proc. of the Australian Water and Wastewater Association 16th Federal Convention*. Sydney, Australia: Australian Water and Wastewater Association.

Camper, A.K. 2003. Bacterial Regrowth Modeling. Presented at Predictive Models for Water Quality in Distribution Systems Workshop, Voorhees, April 1–3, 2003.

Cesario, A.L., J.R. Kroon, and W. Grayman. 1996. New Perspectives on Calibration of Treated Water Distribution System Models. In *Proc. of 1996 AWWA Annual Conference*. Denver, Colo.: AWWA.

Chambers, V.K., J.D. Creasy, and J.S. Joy. 1995. Modelling Free and Total Chlorine Decay in Potable Water Distribution Systems. *Journal of Water Science Research and Technology—Aqua*, 44(2):60–69.

Characklis, W.G. 1981. Fouling Biofilm Development: A Process Analysis. *Biotechnology and Bioengineering*, 23:1923–1960.

Clark, R.M. 1991. Water Quality Modeling in Distribution Systems: Overview and Conference Summary. In *Proc. of AwwaRF and USEPA Conference on Water Quality Modeling in Distribution Systems*. Denver, Colo.: AWWA.

Clark, R.M., and W.M. Grayman. 1998. *Modeling Water Quality in Drinking Water Distribution Systems*. Denver, Colo.: AWWA.

Clark, R.M., W.M. Grayman, J.A. Goodrich, R.A. Deininger, and A.F. Hess. 1991. Field-Testing Distribution Water Quality Models. *Journal of the American Water Works Association*, 183(7):67–75.

Clark, R.M., W.M. Grayman, and R.M. Males. 1988. Contaminant Propagation in Distribution Systems. *Journal of Environmental Engineering*, 114(4):929–943.

Clark, R.M., L.A. Rossman, and L.J. Wymer. 1995. Modeling Distribution System Water Quality: Regulatory Implications. *Journal of Water Resources Planning and Management*, 121(6):423–428.

Clark, R.M., and M. Sivaganesan. 1998. Predicting Chlorine Residual and Formation of TTHMs in Drinking Water. *Journal of Environmental Engineering*, (December 1998):1203–1210.

Clark, R.M., and M. Sivaganesan. 2002. Predicting Chlorine Residuals in Drinking Water: A Second Order Model. *Journal of Water Resources Planning and Management*, 128(2):1–10.

Clark, R.M., G. Smalley, J.A. Goodrich, R. Tull, L.A. Rossman, J.J. Vasconcelos, and P.F. Boulos. 1994. Managing Water Quality in Distribution Systems: Simulating TTHM and Chlorine Residual. *Journal of Water Science Research and Technology—Aqua*, 43(4):182–191.

Clarke, G.P., A. Kashti, A. McDonald, and P. Williamson. 1997. Estimating Small Area Demand for Water: A New Methodology. *Journal of the Chartered Institute of Water and Environmental Management*, 11(3):186–192.

Clement, J., M. Hayes, P. Sarin, V.L. Snoeyink, W.M. Kriven, J. Bebee, K. Jim, M. Beckett, G.J. Kirmeyer, and G. Pierson. 2002. *Development of Red Water Control Strategies*. Denver, Colo.: AwwaRF and AWWA.

Coulson, J.M., and J.F. Richardson. 1964. *Chemical Engineering*. 2nd ed. Oxford, England: Pergamon Press.

Cubillo, F., and M. Mercier. 1993. An Approach to Water-Quality Modelling in the Paris and Madrid Distribution Networks. *Water Supply*, 11(3–4):377.

Deininger, R.A., R.M. Clark, A.F. Hess, and E.V. Bernstam. 1992. Animation and Visualization of Water Quality in Distribution Systems. *Journal of the American Water Works Association*, 184(7):48–52.

De Rosa, S., and S. Miller. 1994. *Water Quality in Distribution—Final Report*. Report UM 1430. Common Interest Programme, Reference U-0509. Swindon, England: Water Research Centre.

Desjardins, A., V. Gauthier, E. Mainville, R. Millette, and C. Marche. 2000. Determination of Residence Time in Water Distribution Systems by Signal Processing. In *Proc. of the Annual AWWA Water Quality Technology Conference*. Denver, Colo.: AWWA.

Economic and Engineering Services Inc. 1990. *Lead Control Strategies*. Denver, Colo.: AwwaRF and AWWA.

Edwards, K., and L. Martin. 1995. A Methodology for Surveying Domestic Water Consumption. *Journal of the Chartered Institute of Water and Environmental Management*, 9(5):477–488.

Engineering Computer Applications Committee. 1999. Calibration Guidelines for Water Distribution System Modeling. In *Proc. of AWWA Information Management and Technology Conference*. Denver, Colo.: AWWA.

Fair, G.M., and J.C. Geyer. 1954. Disinfection, Destruction of Aquatic Growths & Removal of Odors and Tastes. In *Water Supply and Waste-Water Disposal*. London, England: John Wiley and Sons.

Feben, D., and M.J. Taras. 1951. Studies on Chlorine Demand Constants. *Journal of the American Water Works Association*, 43(11):922.

Gagnon, G.A., P.J. Ollos, and P.M. Huck. 1997. Modelling BOM Utilisation and Biofilm Growth in Distribution Systems: Review and Identification or Research Needs. *Journal of Water Science Research and Technology—Aqua*, 46(1):165–180.

Gale, P., R. Lacey, G. Stanfield, and D. Holt. 1997. The Development and Application of Coliform Predictive Models for the UK Water Industry. *Journal of Water Science Research and Technology—Aqua*, 46(4):185–195.

Gatel, D., M. Mercier, C. Volk, and J-C. Joret. 1995. Control of Bacterial Regrowth in Distribution Systems: Influence of BDOC and Chlorine Residual. *Water Supply*, 13(2):275–290.

Gauthier, V., C. Rosin, L. Mathieu, J.-M. Portal, J.-C. Block, P. Chaix, and D. Gatel. 1996. Characterisation of the Loose Deposits in Drinking Water Distribution Systems. In *Proc. of the AWWA Water Quality Technology Conference*. Denver, Colo.: AWWA.

Geldreich, E.E. 1991. Investigating the Outbreak in Cabool, Missouri for a Water Supply Connection. In *Proc. of AwwaRF and USEPA Conference on Water Quality Modelling in Distribution Systems*. Denver, Colo.: AWWA.

Gotoh, K. 1989. Chlorine Residual Decreasing Rate Coefficients for Various Pipe Materials. *Water Supply*, 7(2–3):164–172.

Grayman, W.M. 2001. Use of Tracer Studies and Water Quality Models to Calibrate a Network Hydraulic Model. *Current Methods*, 1(1):38–42.

Grayman, W.M., R.M. Clark, and R.M. Males. 1988. Modeling Distribution-System Water Quality: Dynamic Approach. *Journal of Water Resources Planning and Management*, 114(3):295–312.

Grayman, W.M., L.A. Rossman, C. Arnold, R.A. Deininger, C. Smith, J.F. Smith, and R. Schnipke. 2000. *Water Quality Modelling of Distribution System Storage Facilities*. Denver, Colo.: AwwaRF and AWWA.

Grayman, W.M., L.A. Rossman, M.A. Gill, Y. Li, and D.E. Guastella. [N.d.] Measuring and Modeling Disinfectant Wall Demand in Metallic Pipes. Forthcoming.

Haas, C.N., and J. Joffe. 1994. Disinfection Under Dynamic Conditions: Modification of Hom's Model for Decay. *Environmental Science and Technology*, 28:1367–1369.

Haas, C.N., J. Joffe, U. Anmangandla, J.G. Jacangelo, and M. Heath. 1996. Water Quality and Disinfection Kinetics. *Journal of the American Water Works Association*, 88(3):95–103.

Haas, C.N., and S.B. Karra. 1984. Kinetics of Wastewater Chlorine Demand Exertion. *Journal of the Water Pollution Control Federation*, 56(2):170–173.

Halhal, D., G.A. Walters, D. Ouazar, and D.A. Savic. 1997. Water Network Rehabillitation With Structured Messy Genetic Algorithm. *Journal of Water Resources Planning and Management*, 123(3):137–146.

Hua, F., J.R. West, R.A. Barker, and C.F. Forster. 1999. Modelling of Chlorine Decay in Municipal Water Supplies. *Water Research*, 33(12):2747–2756.

Jadas-Hécart, A., A. El Moher, M. Stitou, P. Bouillot, and B. Legube. 1992. The Chlorine Demand of a Treated Water. *Water Research*, 26(8):1073–1084.

Kapelan, Z.S., D.A. Savic, and G.A. Walters. 2003. Multiobjective Sampling Design for Water Distribution Model Calibration. *Journal of Water Resources Planning and Management*, 129(6):466–479.

Kirmeyer, G.J., M. Friedman, K. Martel, D. Howie, M. LeChevallier, M. Abbaszadegan, M. Karim, J. Funk, and J. Harbour. 2001. *Pathogen Intrusion Into the Distribution System*. Denver, Colo.: AwwaRF and AWWA.

Laurent, P., P. Servais, M. Prevost, D. Gatel, and B. Clement. 1997. SANCHO Model of Microbiological Water Quality in Distribution Systems. *Journal of the American Water Works Association*, 89(7):92–103.

LeChevallier, M.W., W. Schulz, and R.G. Lee. 1991. Bacterial Nutrients in Drinking Water. *Applied and Environmental Microbiology*, 57(3):857–862.

Lin, S. 2001. *Water and Wastewater Calculations Manual*. New York: McGraw-Hill.

Liou, C.P., and J.R. Kroon. 1987. Modeling the Propagation of Waterborne Substances in Networks. *Journal of the American Water Works Association*, 79:(11)54–58.

Lu, C., P. Biswas, and R.M. Clark. 1995. Simultaneous Transport of Substrates, Disinfectants and Mircoorganisms in Water Pipes. *Water Research*, 29(3):881–894.

Macke, E. 1982. About Sedimentation at Low Concentrations in Partly Filled Pipes. *Mitteilungen, Leichtweiss–Institut für Wasserbau de Technischen Universität Braunshweig, Heft*, 76:1–151.

Males, R.M., W.M. Grayman, and R.M. Clark. 1988. Modelling Water Quality in Distribution Systems. *Journal of Water Resources Planning and Management*, 114(2):197–209.

Marshall, G.P. 2001. *Understanding and Preventing Discoloured Water.* London, England: United Kingdom Water Industry Research.

Marshall, K.C. 1976. *Interfaces in Microbial Ecology*. Cambridge, Mass., and London, England: Harvard University Press.

Mary-Dile, V., D. Gatel, P. Servais, C. Elain, and J. Cavard. 2000. Developing a Water Quality Model for the Distribution System in the Suburbs of Paris, France. In *Proc. of the AWWA Annual Water Quality Technology Conference*. Denver, Colo.: AWWA.

Mau, R.E., P.F. Boulos, R.M. Clark, W.M. Grayman, R.J. Tekippe, and R.R. Trussell. 1995. Explicit Mathematical Model of Distribution Storage Water Quality. *Journal of Hydraulic Engineering*, 121(10):699–709.

May, R.W.P. 1994. Appendix 2, Section A2.2.9, May (1994). In *Design of Sewers to Control Sediment Problems*. CIRIA Report 141. London, England: CIRIA.

Mayer, P.W., W.B. DeOreo, E.M. Opitz, J.C. Kiefer, W.Y. Davies, B. Dziegielewski, and J.O. Nelson. 1999. *Residential End Uses of Water*. Denver, Colo.: AwwaRF and AWWA.

McMath, S.M., and R.J. Casey. 2000. The Use of Network Modelling Techniques to Resolve Water Quality Issues in Distribution Systems In *Proc. of the AWWA Annual Water Quality Technology Conference*. Denver, Colo.: AWWA.

Metcalf & Eddy. 1991. Biological Unit Processes. In *Wastewater Engineering: Treatment Disposal and Reuse*. 3rd ed. Singapore: McGraw-Hill.

Munvalli, G.R., and M.S. Mohan Kumar. 2003. Optimal Scheduling of Multiple Chlorine Sources in Water Distribution Systems. *Journal of Water Resources Planning and Management*, 129(6):493–504.

NanCIE (Centre International de l'Eau de Nancy). 1998. *Chlorine and Monochloramine Demand of Materials for Drinking Water Distribution Systems*. London, England: United Kingdom Water Industry Research.

Parker, M., J.G. Thompson, R.R. Reynolds Jr., and M.D. Smith. 1995. Use and Misuse of Complex Models: Examples From Water Demand Management. *Water Resources Bulletin*, 31(2):257–263.

Pernitsky, D.J., G.R. Finch, and P.M. Huck. 1995. Disinfection Kinetics of Heterotrophic Plate Count Bacteria in Biologically Treated Potable Water. *Water Research*, 29(5):1235–1241.

Piriou, Ph., S. Dukan, Y. Levi, and P.A. Jarrige. 1997. Prevention of Bacterial Growth in Drinking Water Distribution Systems. *Water Science and Technology*, 35(11–12):283–287.

Powell, J.C. 1998. Modelling Chlorine in Water Distribution Networks. Ph.D. diss. The University of Birmingham, Birmingham, England.

Powell, J.C., N.B. Hallam, J.R. West, C.F. Forster, and J. Simms. 2000a. Factors Which Control Bulk Chlorine Decay Rates. *Water Research*, 34(1):117–126.

Powell, J.C., J.R. West, N.B. Hallam, C.F. Forster, and J. Simms. 2000b. Performance of Various Kinetic Models for Chlorine Decay. *Journal of Water Resources Planning and Management*, 126(1):13–20.

Powell, J.C., J.R. West, N.B. Hallam, and J. Simms. 1999. Durability and Maintenance of Chlorine Network Models. In *Proc. of AWWA Information Management and Technology Conference*. Denver, Colo.: AWWA.

Ransome, M.E., T.N. Whitmore, and J. Easton. 1995. Practical Implications of Disinfection. *Water Supply*, 13(2):131–142.

Reckhow, K.H., J.T. Clements, and R.C. Dodd. 1990. Statistical Evaluation of Mechanistic Water-Quality Models. *Journal of Environmental Engineering*, 116(2):250–268.

Reddy, L.S., L.E. Ormsbee, and D.J. Wood. 1995. Time-Averaging Water Quality Assessment. *Journal of the American Water Works Association*, 187(7):64–73.

Rodriguez, M.J., J. Milot, and J.B. Sérodes. 2003. Predicting Trihalomethane Formation in Chlorinated Waters Using Multivariate Regression and Neural Networks. *Journal of Water Supply: Research and Technology—Aqua*, 52(3):199–215.

Rodriguez, M.J., and J.B. Sérodes. 1994. Development of Neural Net-Based Models for Water Quality Managemnet and Prediction in Distribution Systems. In *Proc. of AWWA Conference on Computers in the Water Industry.* Denver, Colo.: AWWA.

Rodriguez, M.J., J.R. West, J. Powell, and J.B. Sérodes. 1997. Application of Two Approaches to Model Chlorine Residuals in Severn Trent Water Ltd (STW) Distribution Systems. *Water Science and Technology*, 36(5):317–324.

Rossman, L.A. 1994. *EPANET Users Manual*. Version 1.1. Cincinnati, Ohio: USEPA.

Rossman, L.A., and P.F. Boulos. 1996. Numerical Methods for Modeling Water Quality in Distribution Systems: A Comparison. *Journal of Water Resources Planning and Management*, 122(2):137–146.

Rossman, L.A., P.F. Boulos, and T. Altman. 1993. Discrete Volume-Element Method for Network Water-Quality Models. *Journal of Water Resources Planning and Management*, 119(5):505–517.

Rossman, L.A., R.M. Clark, and W.M. Grayman. 1994. Modeling Chlorine Residual in Drinking-Water Distribution Systems. *Journal of Environmental Engineering*, 120(4):803–820.

Safe Drinking Water Committee. 1982. Biological Quality of Water in the Distribution System. In *Drinking Water and Health*. Vol. 4. Washington, D.C.: National Academy Press.

Sarin, P., J. Bebee, M.A. Beckett, K.K. Jim, D.A. Lytle, J.A. Clement, W.M. Kriven, and V.L. Snoeyink. 2000. Release of Iron from Corroded Iron/Steel Pipes in Water Distribution Systems—Effect of Stagnation Time. In *Proc. of International Distribution Research Symposium*. Denver, Colo.: AWWA.

Sarin, P., V.L. Snoeyink, J. Bebee, W.M. Kriven, and J.A. Clement. 2001. Physico-Chemical Characteristics of Corrosion Scales in Old Iron Pipes. *Water Research*, 35(12)2961–2969.

Schock, M.R. 1989. Understanding Corrosion Control Strategies for Lead. *Journal of the American Water Works Association*, 81(7):88–100.

Schock, M.R., and J.R. Clement. 1998. Lead and Copper Control With Non-Zinc Orthophosphate. *Journal of the New England Water Works Association*, 112(1):20.

Seidal, C., M.M. Frey, R. Hund, N. Graziano, B. Takeda, B. Wischer, and J. McEnroe. 2002. The IDSE: An Alternative Approach Using Distribution System Models. In *Proc. of the 2002 AWWA Annual Conference*. Denver, Colo.: AWWA.

Servais, P., P. Laurent, G. Billen, and D. Gatel. 1995. Development of a Model of BDOC and Bacterial Biomass Fluctuations in Distribution Systems. *Revue des Sciences de L'Eau*, 8(4):427–462.

Sharp, W.W., J. Pfeffer, and M. Morgan. 1991. Insitu Chlorine Decay Rate Testing. In *Proc. of* AwwaRF and USEPA Conference on Water Quality Modeling in Distribution Systems. Denver, Colo.: AWWA.

Skipworth, P.J., J. Machell, and A.J. Saul. 2002. Empirical Travel Time Estimation in a Distribution Network. *Water and Maritime Engineers*, 154(1):41–49.

Strafaci, A.M. 2001. Genetic Algorithms in Water Resources Engineering. *Current Methods*, 1(1):119–123.

Teefy, S. 1996. *Tracer Studies in Water Treatment Facilities: A Protocol and Case Studies*. Denver, Colo.: AwwaRF and AWWA.

Thommann, R.V. 1982. Verification of Water Quality Models. *Journal of Environmental Engineering*, 108(5):923–940.

USEPA (U.S. Environmental Protection Agency). 2002. EPANET 2.0: A Water Distribution Hydraulic and Water Quality Simulation Program for Personal Computers. [Online]. Available: <http://www.epa.gov/ORD/NRMRL/wswrd/epanet.html>. [cited January 6, 2003]

USEPA (U.S. Environmental Protection Agency). 2003. *The Stage 2 DBPR Initial Distribution System Evaluation Guidance Manual*, July 2003, Draft [Online]. Available: <http://www.epa.gov/safewater/stage2/guides.html>. [cited November 11, 2003]

Valentine, R.L., K. Ozekin, and P.J. Vikesland. 1998. *Chloramine Decomposition in Distribution Systems and Model Waters*. Denver, Colo.: AwwaRF and AWWA.

Valentine, R.L., and P.J. Vikesland. 2000. *Role of the Pipe-Water Interface in Distinfectant By-Product Formation and Disinfectant Loss*. Denver, Colo.: AwwaRF and AWWA.

van der Kooij, D., J.H.M. van Lieverloo, P. Gale, and G. Stanfield. 2002. *Distributing Drinking Water With Low or Zero Disinfectant Residual*. London, England: United Kingdom Water Industry Research.

Vasconcelos, J.J., P.F. Boulos, W.M. Grayman, L. Kiene, O. Wable, P. Biswas, A. Bhari, L.A. Rossman, R.M. Clark, and J.A. Goodrich. 1996. *Characterization and Modeling of Chlorine Decay in Distribution Systems*. Denver, Colo.: AwwaRF and AWWA.

Vasconcelos, J.J., L.A. Rossman, W.A. Grayman, P.F. Boulos, and R.M. Clark. 1997. Kinetics of Chlorine Decay. *Journal of the American Water Works Association*, 89(7):54–65.

Vítovský, J.P., J.A. Ligget, A.R. Simpson, and M.F. Lambert. 2003. Optimal Measurement Site Locations for Inverse Transient Analysis in Pipe Networks. *Journal of Water Resources Planning and Management*, 129(6):480–491.

Volk, C., and Joret, J.C. 1994. Parametres Predictifs de L'Apparition des Coliformes dans les Reseaux de Distribution D'Eau D'Alimentation. *Rev. Sci de l'Eau*, 7:145–166.

Walski, T., D.V. Chase, and D. Savic. 2001. *Water Distribution Modeling*. Waterbury, Conn.: Haestad Press.

Walski, T., D.V. Chase, D. Savic, W.M. Grayman, S. Beckwith, and E. Koelle. 2003. *Advanced Water Distribution Modeling and Management*. Waterbury, Conn.: Haestad Press.

Westerhoff, P., D. Reckhow, G. Amy, and Z. Chowdhury. 2002. *Mechanistic-Based Disinfectant and Disinfectant By-Product Models*. USEPA Grant #R 826831-01-0, Final Report [Online]. Available: <http://ceaspub.eas.asu.edu/pwest/USEPA-CL2-DBP-2002.htm>. [cited January 6, 2003]

Woodward, C.A., C.T. Ta, D. Holt, and J. Colbourne. 1996. Relationships Between Observed Monochloramine Decay Rates and Other Physical and Chemical Parameters in a Large Scale Experimental Pipe System. Presented at meeting of Institution of Water and Environmental Management South Eastern Branch, Redhill, Surrey, England, 17 January 1996.

WRc (Water Research Centre Ltd.). 1989. *Network Analysis—A Code of Practice*. Swindon, England: WRc.

WRc (Water Research Centre Ltd.). 1997. *Predictive Statistical Analysis of Coliform Occurrences*. London, England: United Kingdom Water Industry Research.

Wu, Z.Y., P.F. Boulos, C.H. Orr, and J.J. Ro. 2001. Using Genetic Algorithms to Rehabilitate Distribution Systems. *Journal of the American Water Works Association*, 93(11):74–85.

Zhang, G.R., L. Kiene, O. Wable, U.S. Chan, and J.P. Duguet. 1992. Modelling of Chlorine Residual in the Water Distribution Network of Macao. *Environmental Technology*, 13(10):937–946.

ABBREVIATIONS

ANN	artificial neural network
AOC	assimilable organic carbon
ATP	adenosine triphosphate
AWWA	American Water Works Association
AwwaRF	Awwa Research Foundation
BDOC	biodegradable organic carbon
$[Br]$	bromine concentration
°C	degrees Celsius
C	concentration
C_0	initial concentration
CCL	Contaminant Candidate List
CFD	computational fluid dynamics
cfu/mL	colony forming units per milliliter
C_L	limiting concentration
$[Cl_2]$	chlorine dose
CSTR	continuously stirred tank reactor
CT	product of concentration and contact time
C_v	sediment transport capacity
d	pipe diameter
D	molecular diffusivity
d^{-1}	per day
DBP	disinfection by-product
DBPR	Disinfectants and Disinfectant By-Product Rule
DMA	district meter area (leakage metering zone)
d_{p50}	particle size which is larger than 50% by mass of all the particles in a graded sample
E	activation energy
EPS	extended period simulation
f	curve fitting parameter representing offset of k versus $[TOC]$
F	frequency factor
f{}	function of parameter enclosed in brackets
g	acceleration due to gravity
GA	genetic algorithm
GAC	granular activated carbon
GIS	geographical information system
GLM	generalized linear modeling

H	concentration of reactant
H_0	initial concentration of reactant
h^{-1}	per hour
HAA	haloacetic acid
HPC	heterotrophic plate count
IDSE	initial distribution system evaluation
J	joule
J/mol	joules per mole
J/mol °C	joules per mole and degrees Celsius rise in temperature
k	reaction rate constant
k^*	pseudo first order reaction rate constant
k_b	bulk water decay coefficient
k_f	mass transfer coefficient
K_S	substrate concentration at half the maximum growth rate
k_{vcsc}	Valentine chloramine stability index
k_w	wall reactivity coefficient
k_w*	wall decay coefficient
L	length
L	liter
LRAA	local running annual average
λ	Darcy-Weisbach friction factor
m	Hom disinfection kinetic parameter (variable units)
MCL	maximum contaminant level
mg	milligram
mg/L	milligrams per liter
mg PO_4/L	milligrams as PO_4 per liter
mL	milliliter
mm	millimeter
mol	mole
MWRA	Massachusetts Water Resource Authority
μ'	specific growth rate
μ'_m	maximum specific growth rate
μg	microgram
μg/L	micrograms per liter
μm	micrometer
n	reaction order
n'	coefficient of dilution
N	number of surviving microorganisms at time t
N_0	number of microorganisms at time zero
NanCIE	Centre International de l'Eau de Nancy

N.d.	no date
NDMA	N-nitrosodimethylamine
$[NH_2Cl]$	monochloramine concentration
$[NH_2Cl]_0$	initial monochloramine concentration
NOM	natural organic matter
ν	kinematic viscosity
p	curve fitting parameter representing gradient of k versus $[TOC]/C_0$
P	probability
PAC	Project Advisory Committee
PVC	polyvinyl chloride
R	ideal gas constant (8.31 J/mol°C)
RAC	Research Advisory Council
R_e	Reynolds number
RFP	request for proposal
s	second
s	specific gravity of the solids particles
S	stoichiometry of reaction
S'	concentration of growth limiting substrate in solution
Sc	Schmidt number
Sh	Sherwood number
SMP	standard monitoring program
SSS	system-specific study
SWTR	Surface Water Treatment Rule
t	time
T	temperature (°C)
TCR	Total Coliform Rule
THM	trihalomethane
$[THM]$	THM concentration at time t
TOC	total organic carbon
$[TOC]$	TOC concentration
u	velocity
UK	United Kingdom
UKWIR	United Kingdom Water Industry Research Ltd.
USA	United States of America
USEPA	United States Environmental Protection Agency
UV	ultraviolet
$[UV]$	ultraviolet absorbance
V_L	limiting velocity for deposition
V_t	velocity at threshold of motion

W	fall velocity of sediment particles
WRc	Water Research Centre Ltd.
WTP	water treatment plant
X_0 to X_7	WTP model fitted coefficients

IWA Publishing's authorised EU representative for General Product Safety Regulations is Diane D'Arras, 15 rue Duret, 75116 Paris, France, e-mail: safety@iwap.co.uk.

Printed and bound by CPI Group (UK) Ltd, Croydon, CR0 4YY
17/09/2026
02218799-0016